D1401276

Force, Motion, and Energy

Uri Haber-Schaim
Reed Cutting
H. Graden Kirksey
Harold A. Pratt
Robert D. Stair

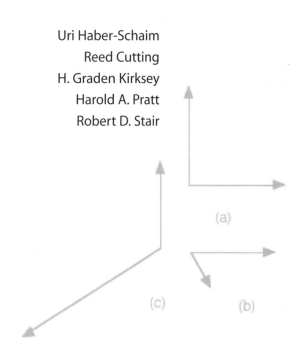

Science Curriculum Inc. • Belmont, MA 02478

Force, Motion, and Energy

Uri Haber-Schaim · Reed Cutting · H. Graden Kirksey ·
Harold A. Pratt · Robert D. Stair

Credits
Editor: Sylvia Gelb
Book Design: SYP Design & Production, Inc.
Production: Andrea Rudner
Photography:
    Benoit Photography: pages 1, 3, 10, 12, 13, 16, 21, 23, 25,
    28, 29, 32, 32, 37, 38, 41, 51, 60, 61, 68, 77, 100, 104, 114,
    115, 122, 123, 127
    Energy, A Sequel to IPS: pages 119, 130, 131, 132, 133
    Peter Gendel: page 113
    Kodanasha Co. Ltd.: page 85
    Martha Svatek: page 47
    PSSC Physics: page 89
    Susan Van Etten: pages 67, 83, 97
Illustrations: Chuck Mackey

Printed in the United States of America
by Quebecor World

ISBN 1–882057–12–0

10 9 8 7 6 5 4 3 2

Preface

The purpose of this program is to give students in eighth- or ninth-grade an understanding of some of the basic ideas of physics, and an insight into the way scientific knowledge is acquired. *Force, Motion, and Energy* (*FM&E*) has been designed as a stand-alone program for a one-semester course. However, since its content complements that of *Introductory Physical Science* (*IPS*), *FM&E* also can be used in a one-year science course that includes the first half of *IPS*. The new program has been field-tested both ways.

Every major concept in *FM&E* is introduced with a laboratory investigation that is a fully integrated part of the program—not an optional add-on. This approach helps students to gain an in-depth understanding of basic principles and their application, as well as useful laboratory skills and the inquiry skills called for by the *National Science Education Standard*s. As students use laboratory data to draw conclusions and formulate generalized principles, they develop their reasoning and problem-solving skills.

FM&E students learn from nature, their text, the teacher, and one another. The teaching strategy employed in this program calls for students to work in teams of two or three and pool their data with the rest of the class to reach a class conclusion. This cooperative learning process requires communication of ideas among the students in written as well as spoken form. The personal laboratory notebooks in which students record their findings become an important supplement to the text.

The selection of content for the program follows a well-developed path, or story line, working toward a carefully selected set of fundamental outcomes called for in the *National Science Education Standard*s as well as most state standards. Because the content is selected for its scientific value and usefulness in future science courses, the book avoids the pitfall of being "a mile wide and an inch deep."

Acknowledgments

The program was piloted in a diverse group of schools and classrooms across the country. Pilot teachers were from both public and private schools. Both eighth- and ninth-grade students were included in the piloting, as well as one select group of seventh-graders.

The individuals listed below provided many useful comments on various components and aspects of the program, including the *Teacher's Guide and Resource Book:*

Peter Brock, Providence Country Day School, Providence, RI
Kathy Colombo, Inglewood Junior High School, Redmond, WA
Heidi Johnson, Inglewood Junior High School, Redmond, WA
Mary Lynn Marden, Neah-Kah High School, Rockaway Beach, OR
Diana Rea, Inglewood Junior High School, Redmond, WA
Janet Steuart, Greco Middle School, Temple Terrace, FL
David Taylor, Triton Regional High School, Byfield, MA

We are especially grateful to the following teachers for their detailed and in-depth feedback, which had a significant impact on the final shape of the program:

Peter Gendel, Isidore Newman School, New Orleans, LA
Anne Prouty, Gilkey Middle School, Portland, OR
Kent Roberts, Colville Junior High School, Colville, WA
Martha Svatek, Nashoba Brooks School, Concord, MA

We are indebted to Angela Carbone, our Coordinator of School Services, who provided a wide range of valuable services in our office during the development and production of the course materials.

Uri Haber-Schaim
Reed Cutting
Graden Kirksey
Harold Pratt
Robert Stair

June 2002

Contents

To the Student

The course you are about to begin will be both challenging and rewarding. It will require effort and will reward you with the joy of understanding important topics in physics. We believe this to be so because this has been the experience of students taking the course before you.

You will be devoting a considerable portion of your class time to doing experiments, finding out first-hand how nature behaves. To arrive at reliable experimental results requires the collection of a large number of data. You will do this by sharing your team's data with the entire class and drawing conclusions jointly. A well-kept notebook will be of great help in organizing your work.

Direct experimenting will not be your only source of information. All through your life you will be learning from the experience of others transmitted through the written word. Being able to read science is a valuable skill. We hope that this course will help you to acquire it.

Finally, knowledge is useful only if you can apply it to new situations. The many problems between sections and at the end of each chapter will help you to sharpen your problem-solving skills and build a solid foundation for future learning.

The Authors

Chapter 1
Forces

1.1 Introduction

From early childhood you have pushed and pulled on various objects. As long as they were not too heavy, you could easily observe the effect of your push or pull. For example, pulling on the handle makes a wagon move. Also, kicking a ball at rest makes the ball move, since a kick is a very short push.

Since a push and a pull produce quite similar results, they have been given a common name—a *force*. Many times you are aware of a force because you produce the push or pull with your own body.

What if you push or pull on something and it does not move? Something must be pushing or pulling just as hard in the opposite direction. An equal force acting in the opposite direction balanced your force.

Not all forces require a contact (touch) to be exerted. As an example, you have observed a magnet pulling or pushing an object without touching it. If nothing is touching an object, how can you tell whether a force is acting on the object? The same rules apply as when there is contact. If the object was at rest and begins to move, a force was acting on it. If the object does not begin to move when a force is applied, then there must be more than one force acting on the object and the forces balance each other.

Tie a string to a pencil and hold it over your desk. When a pencil hangs from a string, the upward pull exerted by the string balances the downward pull, so the pencil does not fall. Now cut the string. The pencil, which was at rest, will move downward even though nothing was touching it. Apparently, a force was acting on the pencil that made it move downward.

In this chapter, you will learn how to measure a number of forces and find out how strong they are under different conditions.

> **1.** **Does a table exert a force on a cup that rests on it? Explain.**

EXPERIMENT
1.2 Weight and the Spring Scale

Have you ever seen a photograph of someone standing next to a big fish that he or she caught? Chances are the fish was hanging from a device that showed its weight. *Weight* is the pull of the earth on an object, and the device that measures it is called a *spring scale*. In this experiment, you will use weight to create calibration marks for a spring scale such as the one shown in Figure 1.1.

For the purpose of this experiment, your unit of force will be the weight of a volume of water specified by your teacher. An empty plastic milk bottle makes a good container for the water.

To make a measurement scale that can easily be removed, attach a strip of masking tape over the two sets of permanent marks on the spring scale. Then hang an empty plastic milk bottle from the spring scale and carefully mark the position of the force indicator on the masking tape. Since there is no water in the bottle, write "0" next to this mark (Figure 1.2).

Use a 250-mL beaker to measure out the amount of water specified for your team. Pour this water into the plastic bottle and again mark the position of the force indicator on the tape. Write "1" next to this mark.

Now measure out the same amount of water that you used the first time and add it to the water already in the milk bottle. Mark the position of the indicator and label it "2." Repeat this procedure until you have labeled up to "5."

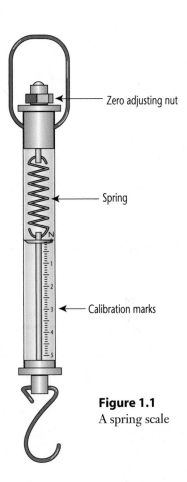

Figure 1.1
A spring scale

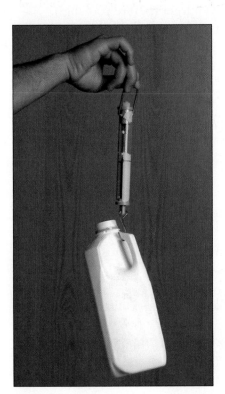

Figure 1.2
The spring scale in Figure 1.1 with an empty bottle and the "0" mark.

The marks you have just made form a measurement scale for your spring scale. The distance between marks is called the measurement *unit* for your scale. In your case, the measurement unit is the weight of the amount of water you added in each step.

- What are the distances between consecutive marks on your tape?
- Does the spring scale stretch by equal distances for equal increases in weight?

Make a smaller or lighter mark exactly halfway between each of the marks you have labeled on the tape.

- What is the value of this division?

Your teacher will provide you with an object. Measure and record its weight, reading the scale to the nearest half unit. Compare the weight you measured for this object with the weights recorded by other students in the class.

- Did all groups report the same weight?

In order for everyone to be able to compare the results of different weight measurements, we need a *standard unit* for weight. The standard for weight, and for all forces in the metric system, is the *newton*. If you look at your spring scale, you will see that one set of calibration marks is labeled "N." This is the abbreviation for "newtons." A newton is about the weight of a small apple or a quarter-pound burger (before cooking!).

To get a better "feel" for the newton, try pulling on the spring scale with a force of 1 N. Then try to estimate the weight of a few objects given to you by your teacher. Weigh these objects and record their weights in newtons.

Carefully peel the masking tape from the spring scale and attach it to a page in your notebook. You will need this tape for a graph you will construct in the next section.

---

2. The spring that you will use in the next section is different from the one you used in this section.

   a. Pull gently on the two springs. Describe the difference between them.

   b. Suppose you hang identical objects on each spring scale. When the objects are at rest, which spring will stretch more? Explain.

   c. Does one spring exert a greater force on the object than the other spring? Explain.

3. What amount of force is shown by each spring scale in Figure A?

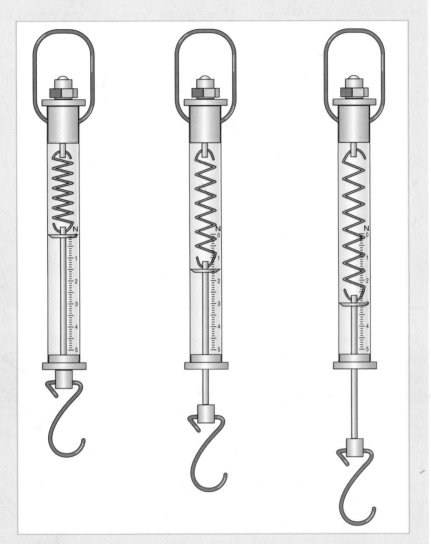

Figure A
For problem 3

4. In the supermarket, weights are given in pounds (abbreviated as "lb"), an old English unit. One pound equals 4.45 N. What is the weight of a 5-lb bag of potatoes in newtons?

5. Suppose that two objects are individually hung from a spring scale and that each object stretches the spring by the same amount. Which of the following properties of the objects—size, shape, weight, color—must be the same and which could be different?

1.3 Hooke's Law: Proportionality

In Experiment 1.2, Weight and the Spring Scale, you made marks on a piece of masking tape to show how much the spring stretched with each increase in weight. If you used the weight of 150 mL of water as your unit of weight, your results may have been similar to those shown in Table 1.1.

Table 1.1 Stretches of a Spring Scale due to Added Weights

Number of weight units added at each step	Distance between marks (mm)
1	7.5
1	7.5
1	7.9
1	7.3
1	7.5

There is another way to present these results. Recall that the zero mark was set at the position of the indicator when the bottle was empty. When one unit of weight was added, the spring stretched 7.5 mm. When the second unit was added, the spring stretched an additional 7.5 mm. Hence, the total stretch for two units of weight was 15.0 mm. Continuing in this way, we can replace Table 1.1 with a table that shows the total stretch produced by the total weight at each stage of the experiment (Table 1.2).

Table 1.2 Total Stretch of a Spring Scale

Total weight (special units)	Total stretch (mm)
0	0
1	7.5
2	15.0
3	22.9
4	30.2
5	37.7

The data in Table 1.2 can also be displayed on a graph (Figure 1.3). The graph is clearly a straight line through the origin. All points on such a graph have a common property: the ratio of their vertical coordinate to their horizontal coordinate is constant. For example, for a total weight of 1.0 special unit, the total stretch is 7.5 mm. This gives a ratio of

$$\frac{7.5 \text{ mm}}{1.0 \text{ special unit}} = 7.5 \text{ mm / (special unit)}.$$

For a total weight of 2.4 special units, the stretch is 18.0 mm. Again, we find a ratio of

$$\frac{18.0 \text{ mm}}{2.4 \text{ special units}} = 7.5 \text{ mm / (special unit)}.$$

When the ratio between two variables is constant, one variable equals the constant value times the other variable. For the graph in Figure 1.3, the relationship between the variables can be expressed as

Stretch (mm) = 7.5 mm/special unit • weight (special units).

Stretch versus Weight

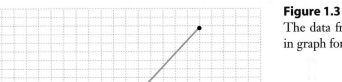

Figure 1.3
The data from Table 1.2 in graph form.

Such a relationship is called a *proportionality*, and the constant is called the *proportionality constant*. (A broader discussion of proportionality is found in Appendix 1 Proportionality.) Note that when the two variables in a proportionality have different units, the unit of the proportionality constant is the ratio of the two units. If the two variables have the same unit, the units cancel and the proportionality constant is simply a number.

The proportionality between the stretch of a spring and the force acting on it is called *Hooke's law*. This law, like other laws of nature, is a generalization based on measurements. Thus, we can have confidence in the law only in the range in which it has been tested. The design of your spring scale prevents you from causing permanent damage to the spring by pulling so hard that you stretch it beyond the range in which Hooke's law applies.

6. Using your own data, make a table similar to Table 1.2. Then graph the total stretch versus total weight. (If you need help in plotting the graph, refer to Appendix 2 Graphing.)

   a. Does your graph represent a proportionality?

   b. If it does, what is the proportionality constant?

7. Examine the newton scales on the 5-N and 10-N spring scales. Make the necessary measurements to draw the stretch-versus-weight graph for each spring scale. What is the proportionality constant for each spring?

8. The perimeter of a square equals four times the length of its side. In other words, the perimeter of a square is proportional to the length of its side.

   a. What is the proportionality constant for this relationship?

   b. Can this proportionality be verified by an experiment? Explain.

   c. Is this relationship limited to squares of a certain size? Explain.

EXPERIMENT
1.4 The Magnetic Force

What happens when two magnets are brought closer to each other? Does the force between them increase or decrease? You have probably played with magnets in the past. You know that whether the magnets are pulling together (*attracting*) or pushing apart (*repelling*), the force between them will be greater when the magnets are closer to each other. But *how* does the magnetic force change as the separation between the magnets changes? If the separation between the magnets is doubled, what happens to the force? In this experiment you will investigate how force and separation are related.

CAUTION: The magnets used in this experiment can damage sensitive electronic devices such as computers and ID cards with magnetically encoded information. Keep the magnets at least 40 cm from such devices.

The apparatus that you will use is shown in Figure 1.4. There are two cylindrical magnets. One is embedded in the base, the other in a dowel glued to a string. The string can be attached to a spring scale. Spacers are placed between the magnets to set the separation at which the force will be measured. All spacers, except the bottom two, are removable, so you can change the separation between the magnets. The lowest two spacers are glued in place to prevent the magnets from coming too close.

To find the separation between the magnets, you must know the thickness of one spacer.

- How will you use the fact that all the spacers are the same thickness to find the most accurate value for the thickness of one spacer?

- What is the thickness of one spacer in millimeters?

Create a data table containing the necessary columns.

CAUTION: Safety glasses must be worn throughout this experiment.

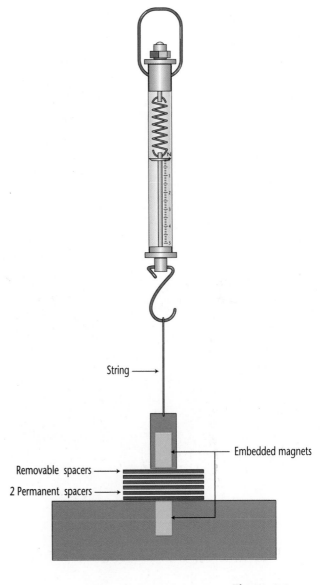

String

Embedded magnets

Removable spacers

2 Permanent spacers

Figure 1.4
Magnetic force apparatus.

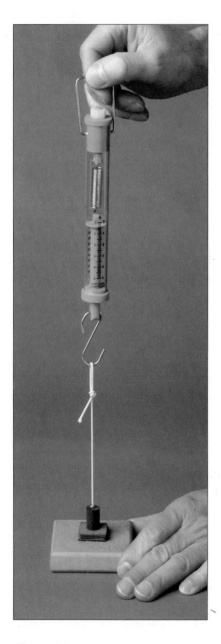

Figure 1.5
Pull straight up on the spring scale to determine the amount of force needed to pull the magnets apart.

A spring scale can be used to measure the magnetic force. But first you must make sure that the spring scale reads only the magnetic force. The weight of the top magnet must be excluded. To do this, use the string to hang the top magnet from the hook on the spring scale. Then adjust the spring scale to read zero by turning the nut at the top of the scale. (Refer to Figure 1.1.) You have now "zeroed" the spring scale for this experiment.

Begin with a total of eight spacers on top of the lower magnet. Carefully lower the top magnet onto the spacers as you hold the string. The hanging magnet will position itself directly above the base magnet. When you begin pulling up on the spring scale, the upper magnet will not move until the spring's pull becomes greater than the pull of the lower magnet (Figure 1.5). To observe the spring scale at the moment the magnets separate, call out the readings of the spring scale as you increase the upward pull. To improve your results, have your lab partner follow closely and repeat the measurement several times. Be sure to record both the separation and the force in your data table.

Now remove spacers from the stack one by one and measure the force required to separate the magnets at each new separation. Make a graph with the magnetic force, in newtons, on the vertical axis and the separation of the magnets, in millimeters, on the horizontal axis.

Draw a smooth curve that fits the data points you have plotted. Remember that your curve does not need to touch all the data points. Do not connect the dots!

- How would you describe your graph to someone who cannot see it?
- According to your graph, what happens to the magnetic force as the separation between the magnets gets larger?

9. Use the graph you constructed for this experiment to answer the following questions.

   a. What do you expect the magnetic force to be at a separation of 4.0 mm? at 9.0 mm?

   b. What do you expect the magnetic force to be at a separation of 50 mm?

   c. What separation is needed between the magnets to provide a magnetic force of 0.40 N? 1.80 N?

EXPERIMENT
1.5 Sliding Friction

To push or pull something with your hand, you must touch the object, that is, you must be in contact with it. The same is true for the force of friction. When you hold a book above a table, there is no friction between the book and the table. For the table to exert a frictional force on the book, the book must be on the table.

Does a horizontal tabletop exert a frictional force on a book resting on it? To find out, push very gently in a horizontal direction on a book resting on a table. If you push gently enough, the book will not move. Yet you can feel that you are pushing on the book. Apparently, a frictional force is balancing your push.

Now push the book with the same force in another direction. The book remains at rest. Apparently, the frictional force also changed direction. It opposes your force just as it did before.

Push the book a little harder. The book still may not move. The frictional force is once again exactly balancing your push. Of course, once you push hard enough, the book will move. The frictional force is a strange force. It does not appear until some other force is present and, up to a point, its strength depends on the strength of that other force.

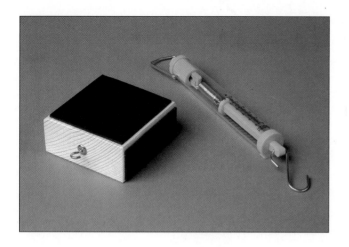

Figure 1.6
Equipment used for studying sliding friction. Nonskid material covers two sides of the friction block.

What factors affect the frictional force between two objects? To help answer this question, you will use a spring scale and a wooden block (Figure 1.6). The block has wide surfaces and narrow surfaces. One wide and one narrow surface are covered with a nonskid material.

Part A

First we will investigate the effect that the kind of surface has on the force of friction.

Clean your lab table thoroughly. Be sure to zero the spring scale while it is in the horizontal position. Begin with the wide covered surface of the friction block in contact with the table (Figure 1.6). Use the spring scale to measure the *smallest* force that will pull the block slowly across the table. Use only enough force to keep the block barely moving. The procedures that you used in the preceding experiment will also be helpful here. Take several readings.

- What is the smallest force that keeps the block moving?
- Should you be concerned about the friction between the spring scale and the table? Why or why not?

Repeat the experiment with the wide uncovered surface of the block in contact with the table.

- What is the smallest force that will keep the block moving now?
- Does the force of friction depend on the types of surfaces that are in contact with each other?

Part B

You can also pull the block when it rests on one of its narrow surfaces. The weight of the block is the same as when it rests on a wide surface. However, the area of contact between the block and the table is now smaller.

With the block resting on one of its narrow surfaces, repeat the two experiments you did in Part A.

- Does the force of friction depend on the area of contact between the block and the table?

10. **Would letting a little air out of the tires of the family car increase the car's traction on an icy road? Why or why not?**

1.6 Friction and Weight

So far, you have investigated whether the force of friction depends on the kind of surface and the contact area between two objects. What about weight? Does the weight of an object affect the friction for the same surfaces?

To study this, we investigated the frictional force on two blocks of different weights, a short (lighter) block and a long (heavier) one (Figure 1.7). Their weights were 1.30 N and 5.05 N. Using the procedure from the preceding experiment, we found that the least force needed to move the lighter block was 0.50 N. For the heavier block, the least force was 2.00 N. The friction was greater for the heavier block than for the lighter one.

We then placed various weights on the lighter block and measured the force of friction for each combination (Figure 1.8). The results are shown

Figure 1.7
Two blocks, a short, lighter one and a long, heavier one, each with nonskid material on the surface with the largest area.

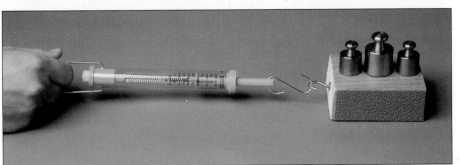

Figure 1.8
Weights have been added to the short, lighter block.

in Table 1.3. These data are graphed in Figure 1.9. Within the range of the data, the force of friction is proportional to weight. The proportionality constant depends on the two surfaces. It is called the *coefficient of friction*.

$$\text{Force of friction} = (\text{coefficient of friction}) \cdot \text{weight}$$

Table 1.3 Data for the investigation of how the frictional force is related to weight. The "Total Weight" is the sum of the weight of the short (lighter) block and the additional weights that were placed on top of it.

Block	Weight of block (N)	Added weight (N)	Total weight (N)	Force of friction (N)
Long	5.05	0.00	5.05	2.00
Short	1.30	0.00	1.30	0.50
Short	1.30	0.49	1.79	0.55
Short	1.30	0.98	2.28	0.90
Short	1.30	1.47	2.77	1.05
Short	1.30	1.96	3.26	1.20
Short	1.30	2.45	3.75	1.40
Short	1.30	2.94	4.24	1.60
Short	1.30	3.43	4.73	1.80
Short	1.30	3.92	5.22	1.90
Short	1.30	4.41	5.71	2.25
Short	1.30	4.90	6.20	2.50

Because both weight and friction are forces, their ratio is a pure number if both are measured in the same units. In Figure 1.9 both forces are expressed in newtons. Hence, the coefficient of friction calculated from the graph is a pure number.

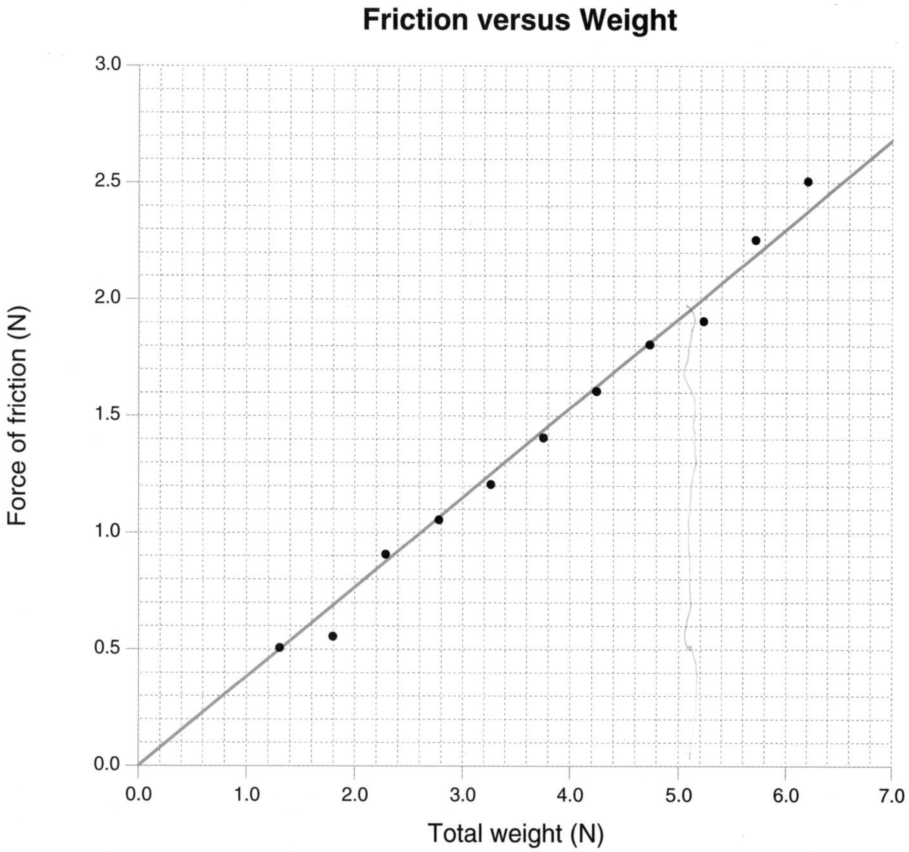

Friction versus Weight

Figure 1.9
The graph of the results shown in Table 1.3.

12. For identical surfaces, how does the force of friction depend on the weight of the object you are trying to move?

13. What is the coefficient of friction for the surfaces used to produce the data in Table 1.3?

14. Why do some motorists keep sand in the trunk of their car for winter driving?

15. Use the graph in Figure 1.9 to answer the following questions.

   a. What would be the frictional force on the short block if its total weight equaled the weight of the long block?

   b. Compare this force with the one measured on the long block.

   c. Are you surprised by your answer to part (b)? Explain.

1.7 Newton's Third Law

In Experiment 1.5, Sliding Friction, when you just began to pull on the block with a spring scale, the block did not move. The frictional force exerted by the table on the block balanced the force exerted by the spring. Did the block also exert a frictional force on the table? If so, how big was that force?

If a frictional force exerted by the pulled block could move the table, we could balance this force using another spring scale. This would tell us how large the frictional force is. Obviously, the table stays put without the force provided by an additional spring scale. So we cannot answer this question.

However, suppose we place the block on a light cart (Figure 1.10). The wheels on the cart, which cannot be seen in the photograph, have very good bearings. As a result, there is very little friction between the cart and the table. If we now pull on the block, the cart will move with the block. To keep the cart at rest, the frictional force exerted by the block on the cart must be balanced by another force. Pulling with the second spring scale provides this balancing force.

Note in Figure 1.10 that the cart is pulled to the left and the block is pulled to the right. Let us call the force exerted by the spring on the cart A, and the force exerted by the other spring on the block, D. Figure 1.10 clearly shows that these two forces are equal:

$$A = D.$$

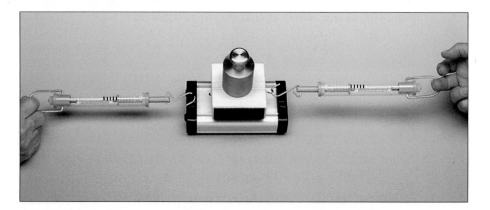

Figure 1.10
A block on a cart. Because of special wheels, there is very little friction between the cart and the table. All three objects are at rest. The cart is pulled to the left and the block is pulled to the right. Note that the readings on both spring scales are the same.

Figure 1.11

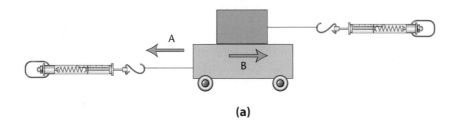

(a)

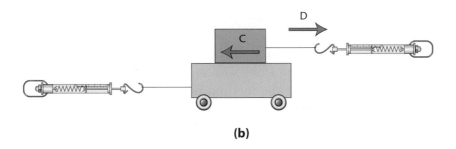

(b)

(b) The forces acting on the block.

Now concentrate on the forces acting on the cart. The spring pulls the cart to the left with a force A. The force of friction exerted by the block pulls the cart to the right with a force B (Figure 1.11(a)). Since the cart is at rest, these two forces must balance each other:

$$A = B.$$

We now apply the same reasoning to the forces acting on the block. The spring pulls the block to the right with a force D. The force of friction exerted by the cart pulls the block to the left with a force C (Figure 1.11(b)). Since the block is at rest, these two forces balance:

$$D = C.$$

We can now substitute B for A and C for D in the first equation, and conclude that

$$B = C.$$

In words, the strength of the frictional force exerted by the block on the cart equals the strength of the frictional force exerted by the cart on the block. These two forces act, of course, in opposite directions.

Now consider another situation. In Experiment 1.4, The Magnetic Force, you measured the force that the lower magnet exerts on the upper magnet. During the experiment, you had to hold the lower magnet in place. Clearly, the upper magnet also exerted a force on the lower magnet. Were the two forces equally strong? You cannot tell on the basis of Experiment 1.4 because you held the lower magnet down with your hand. However, many experiments have shown that the forces that two magnets exert on each other are equal in strength and opposite in direction.

Finally, let us consider another type of force. The weight of a rock on the surface of the earth is the force with which the earth pulls the rock down. This force is called the *gravitational force*. Does the rock pull the earth up with an equal force? No experiment that you can do in the laboratory can detect the effect of the gravitational force that the rock exerts on the earth. To see why this is so, consider the following: If the forces on the friction block did not balance, the block would start moving. If the forces on the upper magnet did not balance, the magnet would start moving. However, in the case of the earth, we would not detect any motion because we would be moving along with the earth.

Nevertheless, studies of the motions of the earth and the moon show that the gravitational forces that the earth and the moon exert on each other are equal in strength. In fact, all forces known today that act between two objects along the line that connects them are equal and opposite. This property of the forces is known as *Newton's third law*.

Newton's third law may appear strange to you. Is it possible that when a heavy truck collides with a small car, the forces acting between them are equal and opposite? "Equal forces" does not mean that the *effects* of those forces are equal. Obviously, the small car may be completely destroyed, whereas the truck may suffer only minor damage. Newton's third law relates only to forces, not to their effects.

16. **On a copy of Figure B, mark all the horizontal forces acting on the block, and all the horizontal forces acting on the cart. Explain which pairs of forces are equal and why.**

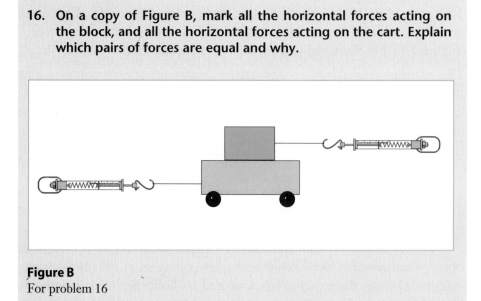

Figure B
For problem 16

17. Consider two different magnets mounted on two carts with low-friction wheels, as shown in Figure C. The carts are at rest. The forces exerted by the two spring scales are equal and opposite. Explain why the forces that the magnets exert on each other are equal and opposite.

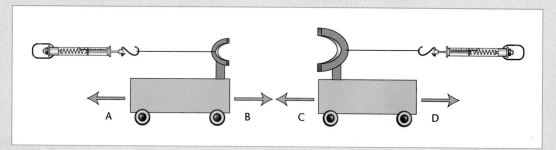

Figure C
For problem 17

FOR REVIEW, APPLICATIONS, AND EXTENSIONS

18. A heavy plate is placed on a vertical spring. The spring is compressed and the plate comes to rest.

    a. What forces are acting on the plate at the moment it is placed on the spring?

    b. What forces are acting on the plate after it has come to rest?

    c. How do the forces in part (b) compare?

19. Most refrigerator doors have a magnetic strip around the edge that holds the door shut. How could you measure the magnetic force holding the door shut? Try it.

20. Not all magnets are permanent. A piece of iron surrounded by a coil carrying an electric current becomes a magnet while the current is turned on. Such a magnet is called an *electromagnet*. In large junkyards, electromagnets are used to lift and move cars.

    a. How much magnetic force is needed to lift a 2,000-lb car?

    b. Express this force in newtons.

    c. Why do you think an electromagnet is used rather than a permanent magnet?

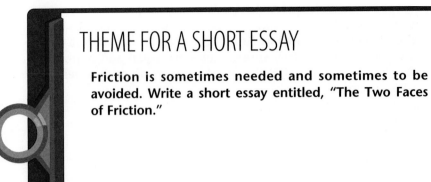

THEME FOR A SHORT ESSAY

Friction is sometimes needed and sometimes to be avoided. Write a short essay entitled, "The Two Faces of Friction."

Chapter 2
Pressure

2.1 Weight and Mass

The spring scales you used in the preceding chapter are not sensitive enough to detect changes of weight within a building. However, using much more sensitive devices, it has been shown that the weight of an object does depend on the elevation at which it is measured (Table 2.1).

Table 2.1 Weight of a Brick at Different Elevations

Location	Elevation (m)	Weight of brick (N)
New York, New York	0	10.000
Denver, Colorado	1,600	9.993
La Paz, Bolivia	3,600	9.977
Top of Mt. Everest	8,848	9.959

All the weight changes indicated in the table are very small. However, suppose that astronauts take a simple spring scale to the moon. What would the scale read for the weight of a brick whose weight on Earth is 10.000 N? On Earth, weight is the pull of the earth on an object. On the moon, the weight of an object is the pull of the moon on it. The answer is that the brick would weigh only 1.61 N on the moon, less than one-sixth of its weight on Earth.

The fact that the weight of an object depends on where it is raises some questions. Suppose you drive from Chicago to Denver, taking a brick with you. Is there less of the brick when you get there? Is there less material in a brick on the moon than in the same brick on Earth? To measure the amount of material in different objects, we need an instrument that will give the same results in Chicago, in Denver, or on the moon. Such an instrument is an *equal-arm balance* (Figure 2.1).

When the beam of an equal-arm balance is horizontal, the weights of the objects on the two pans are equal. When the small container shown in Figure 2.1 was first placed on the left pan, the beam went down on the left side. Adding sufficient weight to the right pan caused the beam to return to a horizontal position. If the balance and the objects it held were then moved to another location, the beam would remain horizontal. The weights of the objects might change, but they would always be equal to each other. Thus, an equal-arm balance does not measure weight, as a spring scale does. We call the property measured by an equal-arm balance *mass*. (We shall also use "mass" as a verb, that is, we shall abbreviate "to find the mass of" as "to mass.")

In Experiment 1.2, Weight and the Spring Scale, you learned that comparing different weights requires the use of common units. We introduced the newton as the standard unit for weight. Similarly, in order to compare measurements on an equal-arm balance, we need a unit for mass. The accepted unit for mass is the *kilogram* (kg). The mass of a platinum cylinder kept under special conditions near Paris, France, is defined to be one kilogram.

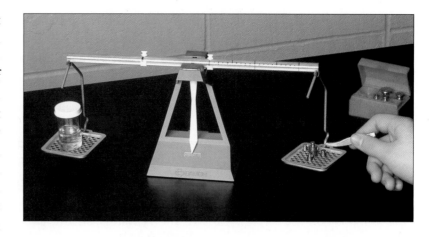

Figure 2.1
An equal-arm balance. A small container is placed on the pan at the left. Standard masses are placed on the pan at the right. The tip of the pointer is at the middle of the scale on the base.

When you buy meat or fruit in the supermarket, you are not really interested in how strongly the earth pulls on your purchase. What interests you is how much meat or fruit you are getting. Yet most stores use spring scales or electronic scales, which operate like spring scales, to measure mass. Such scales can be used to measure mass because the mass of an object is proportional to its weight at any location:

Weight = (proportionality constant) · mass.

If weight is expressed in newtons and mass in kilograms, the constant has units of newtons per kilogram (N/kg).

Since the weight of an object depends on location but its mass does not, the proportionality constant between weight and mass must depend on location. Table 2.2 shows the value of the constant, usually written as g, at the locations included in Table 2.1.

Table 2.2 Values of g at Different Elevations

Location	Elevation (m)	g (N/kg)
New York, New York	0	9.803
Denver, Colorado	1,600	9.796
La Paz, Bolivia	3,600	9.780
Top of Mt. Everest	8,848	9.763

All the values listed in Table 2.2 are very close to 10 N/kg. Thus, for rough measurements on Earth, we need not be concerned with the changes in g at different locations. Look at the spring scale that you used in Experiment 1.2, Weight and the Spring Scale, and you will see two measurement scales on it. One reads "N" for newtons, and the other reads "g" for grams. A *gram* (g) is one-thousandth of a kilogram. (To distinguish between the symbol for gram and that for the proportionality constant, the latter is written in *italics*.)

1. Figure A shows the two scales printed on the spring scale that you used in Experiment 1.2, Weight and the Spring Scale.

   a. What readings on the newton scale correspond to 100, 200, and 500 on the gram scale?

   b. Are the readings on the two scales proportional to one another? Explain.

   c. What is the value of the proportionality constant?

Figure A
For problem 1

2. What is the weight of each of the following food items?
   a. 5.0 kg of apples
   b. 3.5 kg of grapes
   c. a 1.8-kg cantaloupe

3. What is the mass of each of the following food items? (Hint: First convert from pounds to newtons.)
   a. 1.0 lb of meat
   b. 5.0 lb of sugar
   c. 10 lb of flour

4. Suppose a bottle of water for a water cooler weighs 360.00 N in New York City.

   a. Suppose you want the bottle of water to have the same weight in La Paz, Bolivia. Should you add or remove some water?

   b. How much water should you add or remove?

5. Suppose you took a spring balance and some standard masses with you to the moon.

   a. How could you use the spring scale to measure mass on the moon?

   b. How could you use the spring scale to measure weight on the moon?

EXPERIMENT
2.2 Mass, Volume, and Density

All objects have mass. All objects also take up space, so we say that they have *volume*. Are these two quantities related?

To answer this question, you will measure the volume and the mass of four cylinders like those shown in Figure 2.2. Three of the four cylinders are long and thin. The fourth cylinder is shorter and thicker but is made of the same material. This short cylinder has a volume of 1.0 cubic centimeters (cm^3).

Figure 2.2
Four aluminum cylinders. The short cylinder has a volume of 1.0 cm^3.

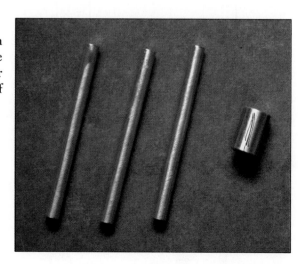

Add some water to a 10-mL graduate. You may find it convenient to use an eyedropper to fill the graduate to a whole number of cubic centimeters for easier reading. Notice that the water surface is curved (Figure 2.3). The curved surface is called a *meniscus*. When reading the volume of a liquid, always read the meniscus at the center, as shown.

Now tilt the graduate and carefully slide the short metal cylinder into the water. Record the new volume.

Caution: To avoid breaking the glass, do not drop the cylinder straight into the graduate when it is in an upright position.

- How does the change in volume compare with the volume of the cylinder?

What you have just done is to measure the volume of the cylinder by a method called *water displacement*. Use this method to measure the volume of each of the three long, thin metal cylinders.

- How do the volumes of these cylinders compare with the volume of the short cylinder?

- Do you expect the mass of these four cylinders to be the same? Why or why not?

Set up a table with columns headed "Mass" and "Volume." Mass any one of the four cylinders. Record its mass next to its volume in your table. Then select any two of the cylinders, determine their total mass, and record it in your table beside the total volume of these cylinders. Repeat

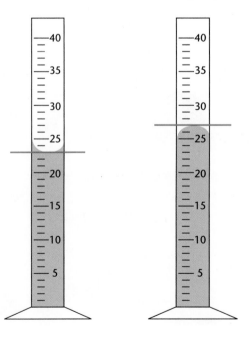

Figure 2.3
A meniscus can curve upward or downward. In either case, read the volume of the liquid at the center of the meniscus.

this procedure for any three of the cylinders and for all four of them. You are now ready to plot a graph of mass as a function of volume.

- Should the origin be a point on your graph?
- Does the graph you drew represent a proportionality? If so, what is the proportionality constant and what are its units?

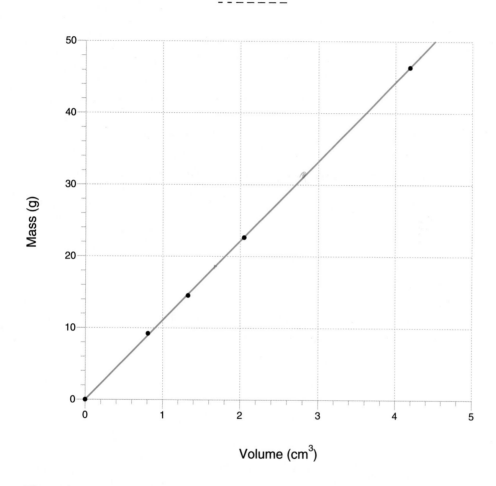

Figure 2.4
A graph of the mass of pieces of lead versus their volume.

Figure 2.4 shows the results of an experiment similar to the one you just did. The relationship between mass and volume is again a proportionality. However, the proportionality constant has a different value. Pieces of lead were used in this experiment instead of the aluminum cylinders that you used. Evidently, the value of the proportionality constant depends on the substance. This constant is called the *density* of the substance:

$$\text{Mass} = \text{density} \cdot \text{volume}$$

Table 2.3 lists the densities for several solids and liquids.

Table 2.3 Densities of Some Solids and Liquids

Substance	Density (g/cm^3)	Substance	Density (g/cm^3)
Gold	19.3	Rock	2.5 to 4.0
Mercury	13.6	Water	1.00
Copper	8.9	Ice	0.92
Iron	7.8	Candle oil	0.76
Magnesium	1.7	Wood	0.6 to 0.9

6. **What is the mass of the given volume of each substance?**
   a. 40.0 cm^3 of mercury   b. 16.0 cm^3 of iron
   c. 7.5 cm^3 of candle oil

7. **A rock has a volume of 220 cm^3. Find the range of possible masses for the rock.**

2.3 Force and Pressure

In Experiment 1.5, Sliding Friction, you found that the force of friction between the block and the table did not depend on the area of contact. This result may have surprised you. A plausible argument to support the opposite result might have been: "When the block slides on a wide surface, a larger area is in contact with the table. Hence, the friction should be greater." On the other hand, the following argument is also plausible: "When the block slides on a narrow surface, its weight is concentrated on a smaller area. Therefore, it presses harder, and the friction should be greater." The two arguments seem to cancel each other. Friction does not depend on area, but it does depend on weight. The dependence of friction on weight rather than area was reinforced in Section 1.6.

There are, however, many situations in which the area over which a force acts is important. The human body can endure large forces provided they are not concentrated on a small area. Someone who can easily lift a 30-lb suitcase would not be able to do so using a thin wire. The wire would cut into the person's hand. Look at the

Figure 2.5
A shoulder bag, with the wider tab on its strap shown.

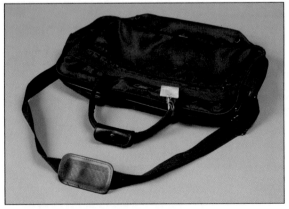

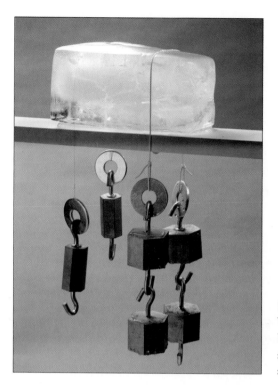

Figure 2.6
Weights hang from both ends of a thin wire and a string draped over a block of ice.

Figure 2.7
A close-up of the setup shown in Figure 2.6.

shoulder bag in Figure 2.5. The strap is quite wide. Yet the part that will rest on the shoulder is made even wider. Widening the strap does not make the bag any lighter. But it does distribute the weight over a larger area on the shoulder, thereby reducing the weight per unit area.

Why does a thumbtack consist of a broad head attached to a thin needle? The force that a thumb exerts on the head of the tack is transmitted through the needle to the bulletin board. For a typical thumbtack, the area of the head is about 70 mm^2, and the cross section of the needle is about 0.9 mm^2. (At the very tip, the area is even smaller.) Applying Newton's third law, the same force is exerted on the thumb and on the bulletin board. But the force on the board acts on an area about 80 times smaller than the force on the thumb does. Hence, the force per unit area is at least 80 times greater on the board than on the thumb. Therefore, the point of the thumbtack penetrates the board but its head does not injure the thumb.

Figures 2.6 and 2.7 show another example in which force per unit area — not simply force alone — is important. A very thin wire and a string, each loaded on both ends with weights, were draped over a block of ice at the same time. The weights hanging from the thin wire exerted a total force of 5 N. The total force on the string was 20 N. After about half an hour, the thin wire cut several millimeters into the ice. The string, with four times as much weight on it, was still sitting on the surface of the ice.

The force per unit area exerted by the wire was many times greater than the force per unit area exerted by the string.

Because force per unit area appears in many situations, it has been given a special name, *pressure*.

$$\text{Pressure} = \frac{\text{force}}{\text{area}}$$

The units used for force and area determine the unit used for pressure. In this chapter, we usually measure forces in newtons and areas in square centimeters. Therefore, the unit used for pressure will be newtons per square centimeter, or N/cm^2.

8. An ordinary brick weighs about 23 N. Its length, width, and height are 21 cm, 9.0 cm, and 6.0 cm, respectively. When the brick rests on a table, what pressure does it exert on the table in each of the three orientations shown in Figure B?

9. Suppose that three bricks of the kind shown in Figure B are stacked on a table as shown in Figure C. What pressure is exerted on the table in each arrangement?

Figure B
For problem 8

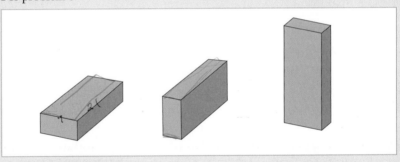

Figure C
For problem 9

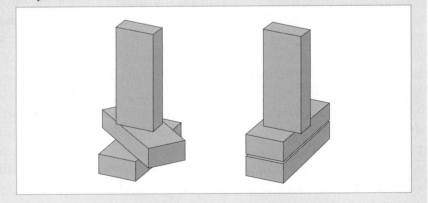

10. To promote its services, a leading hospital in Boston, MA, placed an ad in a newspaper. To catch readers' attention, the ad had a drawing of a woman wearing high-heeled shoes. The caption read: "Did you know that wearing high heels puts more than seven times your body weight directly on the ball of your foot?" Explain what is wrong with the ad, and describe how you would correct it.

11. Use the data in the table to answer the questions.

    a. What is the pressure on the feet of each animal?

    b. What is the ratio of the weight of each of the other animals to the weight of the cat?

    c. What is the ratio of the pressure on the feet of each of the other animals to the pressure on the feet of the cat?

    d. What do you notice about the ratio of the weights compared with the ratio of the pressures?

Animal	Weight (N)	Area of footprint (cm²)	Pressure on feet (N/cm²)
elephant	64,000	4 × 1,100	?
horse	5,000	4 × 110	?
human	750	2 × 120	?
cat	50	4 × 10	?

EXPERIMENT
2.4 Another Type of Balance

In Section 2.1, Weight and Mass, you learned that when the beam of an equal-arm balance is horizontal, the weights on the two pans are equal (Figure 2.1). That is, equal forces on the pans balance the beam.

Consider now two liquid-filled cylinders of different diameter connected with a tube as shown in Figure 2.8. The liquid in the two cylinders is at the same level. What will happen if you put a weight on the liquid in the small cylinder? The simplest way of putting a weight on the liquid is to add a second liquid that does not mix with the first. As Figure 2.9 shows, some of the first liquid flowed into the large cylinder, and its level is no longer the same in both cylinders.

Figure 2.8
Two connected cylinders of different diameter. The level of the colored water is the same in both.

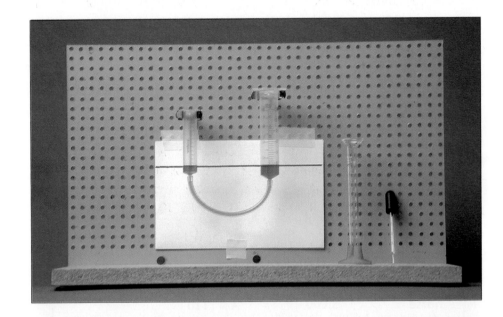

Figure 2.9
After a second liquid is added to the cylinder on the left, the level of the first liquid is different in the two cylinders.

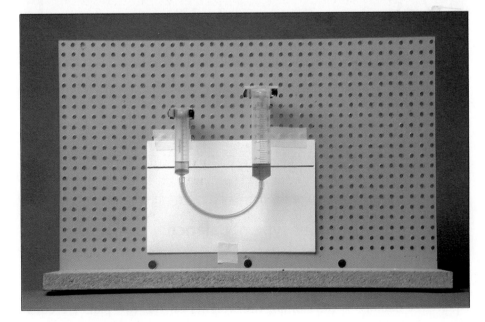

Will putting an equal weight of the second liquid into the large cylinder bring the first liquid to the same level in both cylinders? In other words, do the two connected cylinders act like an equal-arm balance?

The barrels of two syringes of different diameter will serve as the cylinders in this experiment. You will use colored water and candle oil as the two liquids.

Measure the inside diameter of each cylinder. (You will need this information later.) Then set up the apparatus as shown in Figure 2.8. Be sure that the horizontal reference line is about 2 cm above the bottom on each cylinder. Then add colored water until the water levels in both cylinders are at the horizontal line. Add the water slowly to minimize the number of air bubbles that form in the plastic tube.

Any air bubbles that form must be removed. To do this, first remove the small cylinder from its clamp. Then move it up and down to make water flow from one cylinder to the other and push the bubbles out. After all the bubbles have been removed, adjust the background paper so that the horizontal reference line is even with the water levels in *both cylinders*.

You are now ready to begin your investigation by adding 2.0 cm^3 of candle oil from a graduate to the small cylinder. You may want to use an eyedropper to fill the graduate accurately to the 2.0-cm^3 mark.

- Is the level of the colored water in each cylinder at the horizontal reference line?

Will adding 2.0 cm^3 of candle oil to the large cylinder bring the water levels back to the reference line? Try it.

- Are the centers of both water meniscuses back at the reference line?

- Does 2.0 cm^3 of candle oil in the large cylinder balance 2.0 cm^3 in the small cylinder? If not, to which cylinder must oil be added to restore balance?

Add more oil to the appropriate cylinder until the water levels are balanced at the horizontal reference line. This is easily done by first filling your 10-cm^3 graduate with 10.0 cm^3 of oil. Then, carefully pour oil from the graduate into the cylinder until the water levels balance. If you add too much oil, use an eyedropper to remove oil from the cylinder and return it to the graduate. When the water levels balance, you can record the volume of oil remaining in the graduate and find out how much you added to the large cylinder.

- What additional volume of candle oil did you add from the graduate in order to balance the water levels?

- What total volume of oil in the large cylinder balanced 2.0 cm^3 of oil in the small cylinder?

- Using the density of candle oil from Table 2.3, find the mass of oil in each cylinder when the water levels are balanced.
- What is the weight of candle oil in each cylinder when the water levels are balanced?

Earlier you measured the diameters of both cylinders. Remembering that the radius of a circle equals half the diameter, you can find the base area of the oil in each cylinder with this formula:

$$\text{Area of circle} = \pi \cdot (\text{radius})^2$$

Now you can calculate the pressure of the candle oil on the top surface of the water in both cylinders when the water levels are balanced.

- How does the pressure on the water in the large cylinder compare with the pressure on the water in the small cylinder when the water levels are balanced?
- How do the heights of the oil columns compare?

Add another 1.0 cm³ of candle oil to the small cylinder, and balance it with additional oil in the large cylinder.

- How does the total pressure on the water in the large cylinder compare with the pressure on the water in the small cylinder?

12. Tom placed 2.0 cm³ of oil on top of the water in both the small and large cylinders used in Experiment 2.4.

 a. Would the oil layers in each cylinder have the same thickness? Why or why not?

 b. Would the water levels be balanced? Why or why not?

13. Nina placed 2.0 cm³ of oil on top of the water in the same large and small cylinders that you used in Experiment 2.4. From a graduate filled to the 10.0 cm³ mark, she added more oil to the large cylinder in order to balance the water levels. When the levels balanced, the graduate still contained 8.3 cm³ of oil.

 a. What volume of oil did Nina add to the large cylinder to balance 2.0 cm³ of oil in the small cylinder?

 b. What is the volume of the oil in the large cylinder?

 c. What is the mass of the oil in the large cylinder?

 d. What is the weight of the oil in each cylinder?

 e. What is the pressure of the oil on the water in each cylinder?

2.5 Pressure in Liquids

Figure 2.10

In the preceding experiment, you learned that two columns of candle oil of equal height exert the same pressure on the water below. The candle oil served only as a convenient way of putting a weight on the water surface in the two cylinders. Without any candle oil, a column of water also pushes on what is below it. The downward pressure on the water below is the weight of the water above divided by the base area of the water column in the cylinder:

$$\text{Pressure} = \frac{\text{weight}}{\text{area}}$$

This relationship is independent of the shape or size of the base. To better visualize the weight per unit area, we consider a volume of water in the shape of a rectangular box (Figure 2.10(a)). The top of the box is the water surface, and its base is at the depth at which we want to calculate the pressure. Thus, Figure 2.10(a) represents the total volume of the water. The container itself is not shown. The weight per unit area equals the weight of a column with a base of 1 cm². One such column is highlighted in Figure 2.10(a).

The weight of this column equals the weight of water per unit volume times the depth of the column (Figure 2.10(b)). Thus, we can express the weight per unit area, or pressure, as:

$$\text{Pressure} = (\text{weight per unit volume}) \cdot \text{depth}$$

The weight per unit volume of various liquids is known but is not listed in tables because weight varies from place to place. However, weight is proportional to mass. Therefore, weight per unit volume is proportional to mass per unit volume with the same proportionality constant:

$$\frac{\text{Weight}}{\text{volume}} = \frac{g \cdot \text{mass}}{\text{volume}} = g \cdot \text{density}.$$

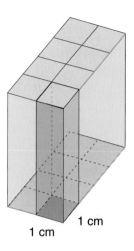

1 cm
1 cm

(a) The small squares all have an area of 1 cm². The weight of the box divided by the base area (8 cm²) is the weight of the water per square centimeter. A column with a base area of 1 cm² is highlighted.

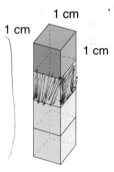

1 cm
1 cm
1 cm

(b) The weight of this column of water equals the weight of water per unit volume times the depth of the water at the level where the pressure is calculated.

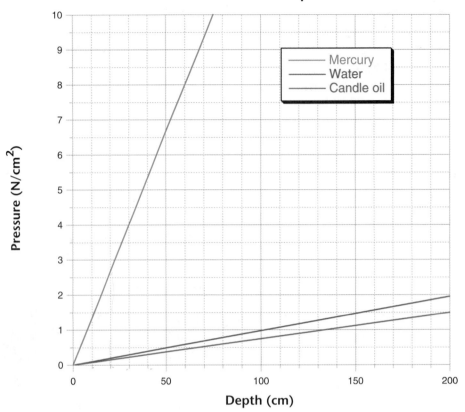

Figure 2.11
A graph of pressure as a function of depth for mercury,
water, and candle oil.

When you need to find the weight per unit volume of a liquid, you can refer to a table like Table 2.3 to find the density. Multiplying the density by g will give the weight per unit volume. The last two formulas were used in calculating the data for Figure 2.11.

Is the pressure in a liquid limited to the downward direction? If you poke a hole in the side of a container below the water level, the water will flow out. This is evidence that a sideways pressure was acting on the water near the wall of the container. As long as the wall pressed back, the water remained at rest. Without the wall pressing back, the water flows out.

How does the pressure at a given depth depend on the direction? To answer this question, we need a device on which the effect of pressure is easy to observe. A tube closed at one end and open at the other end can serve this purpose. When the open end is pushed into a container with water, the water pushes the air back into the tube until the pressure of the air equals the pressure of the water. Three such tubes of equal length are shown in Figure 2.12.

Figure 2.13 shows one of the tubes in a container with water. The two horizontal lines on the container set the depth at which the pressure is measured. The position of the rod is adjusted so that the interface of the air and water is at the same depth as the two lines. Close-ups of the three

Figure 2.12
Three tubes of equal length. All are sealed at the top and open at the bottom. Tube A measures the upward pressure, tube B the sideways pressure, and tube C the downward pressure.

Figure 2.13
One of the tubes inside a vessel filled with water. The interface of the water and the air inside the tube is lined up with the two horizontal black lines on the walls of the vessel.

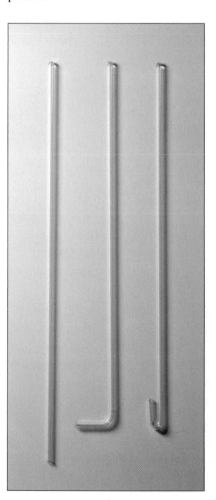

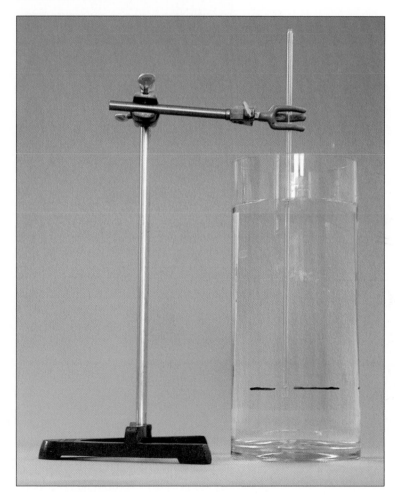

tubes lined up with the two lines are shown in Figure 2.14. In all three directions, the water entered the same distance into the tube. Evidently, the pressure in a liquid does not depend on direction.

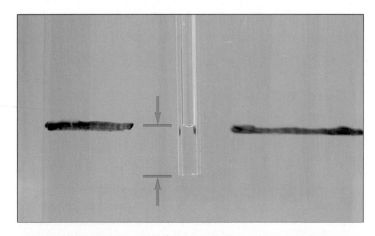

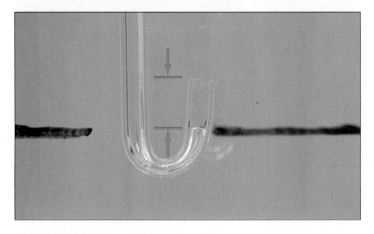

Figure 2.14
The air is pushed back by the same distance, showing that the pressure was the same in all three directions.

14. A cylindrical container with a diameter of 5.0 cm and a height of 40 cm is filled with water.

    a. What is the volume of the water in the container?

    b. What is the mass of the water in the container?

    c. What is the weight of the water in the container?

    d. What is the pressure at the bottom of the container?

    e. What is the pressure at a depth of 20 cm?

    f. What is the pressure at a depth of 10 cm?

15. Suppose the two tall containers shown in Figure D are filled with water. The containers are the same height but have different-sized square bases. How does the pressure at the base of the container on the left compare with the pressure at the base of the container on the right?

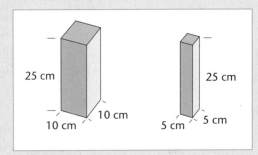

Figure D
For problem 15

16. Three containers are filled to a height of 20 cm with water, candle oil, and mercury, respectively. Using Figure 2.11, find the pressure exerted by each liquid at the bottom of the container.

17. A column of mercury exerts a pressure of 1.0 N/cm^2 on the bottom of a test tube. How tall would a tube full of water have to be for the water to exert this same pressure on the bottom of the tube?

2.6 The Buoyant Force

The pressure exerted in all directions by a liquid depends only on the density of the liquid and on depth. This fact has important consequences.

Consider a solid cylinder held in a container of water by a thin string (Figure 2.15). What forces are acting on the cylinder? First, there is the water above the cylinder pushing it down. This force equals the pressure at the top of the cylinder times the area of the top. Then, there is an upward force on the bottom of the cylinder. This force equals the pressure at the bottom of the cylinder times the same area. Since the pressure is greater at the bot-

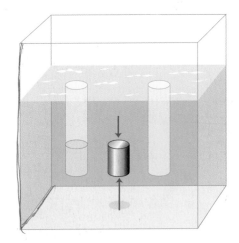

Figure 2.15
A cylinder submerged in a container of water. The volume of the cylinder equals the area of the base (or top) multiplied by the height.

tom of the cylinder than at the top, the combination of these two forces is an upward force, called the *buoyant force*.

There are also sideways forces acting on the cylinder, but these forces balance. For each force pushing on one side of the cylinder, there is an equal force pushing in the opposite direction on the other side of the cylinder. So only the vertical forces acting on the cylinder contribute to the buoyant force.

How large is the buoyant force? The force pushing down on the cylinder is the weight of the column of water above the cylinder. This column is shown in light blue to the left of the solid cylinder in Figure 2.15.

As we have seen in the preceding section, pressure is independent of direction. Also, we know that the areas of the top and bottom of the cylinder are the same. Therefore, the force pushing up on the cylinder is the weight of the column of water shown in light blue to the right of the solid cylinder in Figure 2.15. This column extends from the surface of the water to the base of the solid cylinder.

The difference between these two forces is an upward force equal to the weight of the column shown in yellow in Figure 2.15. The volume of this column of water equals the volume of the submerged cylinder. It is called "the volume of the displaced water."

Of course, we must not forget the weight of the cylinder. The weight of the cylinder is a downward force, whereas the buoyant force is an upward force. If its weight is greater than the buoyant force, the cylinder will sink to the bottom of the container. If its weight is less than the buoyant force, the cylinder will rise to the surface. In effect, the weight of the submerged cylinder is reduced by an amount equal to the weight of the water that it displaces. This law is not limited to cylinders; it holds for objects of any shape. It is called *Archimedes' principle* and has been known since the third century B.C.

Calculating the buoyant force requires special attention to the units used in the relationship

$$\text{Weight} = g \cdot \text{density} \cdot \text{volume}.$$

The proportionality constant g is usually given as 9.8 N/kg. For the kilograms to cancel, we must express the density in kg/cm^3 — not in g/cm^3.

Here is an example. Consider a rod that is 8.0 cm high and has a square base whose sides are each 2.0 cm long. The volume of the rod is:

$$\text{Volume} = 2.0 \text{ cm} \times 2.0 \text{ cm} \times 8.0 \text{ cm} = 32 \text{ cm}^3.$$

This is also the volume of the water displaced by the rod when the rod is completely submerged.

The density of water is 1.0 g/cm^3. Since 1 g = 0.001 kg, the density of water in kg/ cm^3 is 0.0010 kg/ cm^3. Therefore, the weight of the water displaced by the rod in newtons is

Weight of water = 9.8 N/kg × 0.0010 kg/cm^3 × 32 cm^3 = 0.31 N.

18. What is the buoyant force on a cube 10 cm on a side when it is submerged in (a) water and (b) candle oil? (See Table 2.3.)

19. Will any object that sinks in water on Earth also sink in water on the moon? Why or why not?

EXPERIMENT
2.7 Testing a Prediction

One of the characteristics of science is that it allows us to make testable predictions. From the observations of the dependence of pressure on depth, we arrived at Archimedes' principle. This principle makes a testable prediction: The weight of any object submerged in a liquid is reduced by the weight of the liquid that it displaces. In this experiment, you will test this prediction with objects provided by your teacher.

For each of these objects, find the object's weight in air using a spring balance. If the object has a regular shape, make the necessary measurements and calculate its volume. If the object has an irregular shape, you can find its volume by displacement of water.

• What is the weight of each object (in air)?

• What is the volume of each object?

• What is the weight of the water that each object displaces?

• What do you predict will be the weight of each object when it is submerged in water as in Figure 2.16?

• How do your predictions compare with your measurements?

Figure 2.16
A submerged rod being weighed with a spring scale.

2.8 Atmospheric Pressure

Liquids and gases flow. They take on the shape of the container that holds them. Unlike liquids, however, gases fill all the space available to them. A bottle can be half-filled with water, but it cannot be half-filled with air.

Another important difference between liquids and gases is their reaction to pressure. When a piston pushes on a liquid in a closed cylinder, the volume of the liquid barely changes. Liquids are practically *incompressible*. However, pushing on air in a closed cylinder will easily make the air occupy less volume. Gases are *compressible*. The setup shown in Figure 2.13 in Section 2.5 was used to demonstrate that the pressure exerted by about 30 cm of water is enough to produce an observable change in the volume of air in the tube.

Just as water pushes on air, air pushes on water. The atmosphere has weight. Its weight per unit area is *atmospheric pressure*. The effects of atmospheric pressure are easily observed.

When a wide, transparent straw is placed in a glass of water, the water level is the same in the straw as outside the straw. Blowing gently into the straw will lower the water level inside the straw compared with the outside. The pressure inside the straw has been increased. It now takes the atmospheric pressure outside the straw and the additional pressure of a layer of water to balance the pressure inside the straw (Figure 2.17(a)).

Figure 2.17

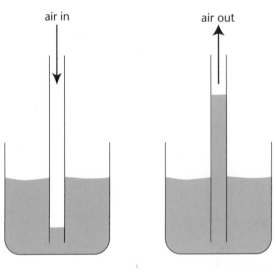

air in

air out

(a) When air is blown in, the water level in the straw drops.

(b) When air is withdrawn, the water level in the straw rises.

Inhaling a little air from the straw will raise the level of the water inside the straw above the level outside. The pressure inside the straw has been reduced. To balance the atmospheric pressure outside now requires an additional column of water inside the straw as shown in Figure 2.17(b). This process happens every time you drink through a straw.

Suppose you could obtain a straw as long as you wanted. How high could you make water rise inside this straw or any long tube? With a long enough tube and a pump to remove all the air from inside the tube, water will rise to a maximum height of about 10.3 m at sea level. No matter how strong the pump is, the water will not rise any higher. At that height, essentially all the air has been evacuated from the tube and the pressure at the bottom of the column of 10.3 m of water is equal to the pressure of the atmosphere.

If we replace the water with mercury, the mercury will rise to 0.760 m or 760 mm. This is a much more convenient height for us to measure (Figure 2.18). If we seal the top of the evacuated tube and place a meter stick next to it, we have made a *barometer* (Figure 2.19). As atmospheric pressure changes, the height of the mercury column also changes. A taller column indicates that the atmosphere is pushing with greater force per unit area on the mercury in the container at the bottom of the barometer. Because we use a barometer, the atmospheric pressure read on the scale is called the *barometric pressure*.

In weather reports, barometric pressure is often expressed in inches of mercury. Sometimes the weather forecaster does not include the unit and says, "The barometric pressure is 29.9." What is really meant is that the barometric pressure is the same as the pressure of a column of mercury with a height of 29.9 in. (or 760 mm).

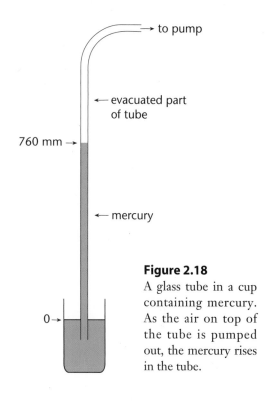

Figure 2.18
A glass tube in a cup containing mercury. As the air on top of the tube is pumped out, the mercury rises in the tube.

Figure 2.19
The upper and lower parts of a mercury barometer. The long middle section is not shown. The scale on the upper part is in both millimeters and inches. Note that the mercury container at the bottom is much wider than the long tube.

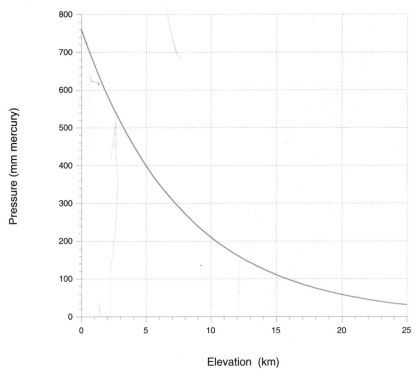

Atmospheric Pressure and Elevation

Figure 2.20
An approximate graph of barometric
pressure as a function of elevation.

An approximate graph of atmospheric pressure as a function of elevation is shown in Figure 2.20.

20. Many aircraft can cruise at an altitude of 40,000 ft (7.6 mi). Use Figure 2.20 and Appendix 3 Conversion of Units to answer the questions.

a. Find the atmospheric pressure at this cruising altitude.

b. Determine the ratio of the air pressure at this cruising altitude to the air pressure at sea level.

21. The atmospheric pressure in Denver is about 4/5 the pressure at sea level. How do you think this would affect the number of homeruns hit during a baseball season?

22. Suppose you hold a 50 cm-by-50 cm tray horizontally.

a. Calculate the approximate weight of the air pushing down on the tray. (A pressure of 760 mm of mercury corresponds to 10.3 N/cm².)

b. Using an average weight of 700 N per person, find the number of people whose combined weight equals the weight of the air over the tray.

c. You could never lift the number of people you determined in part (b), but you can lift the tray. Why?

23. A hollow cube with walls of thick steel is made up of two halves that fit tightly together. The edges of the cube are 50 cm long. Virtually all the air in the cube has been pumped out through a valve.

a. Would you and a friend be capable of pulling the two halves apart?

b. Why must the cube have thick walls?

FOR REVIEW, APPLICATIONS, AND EXTENSIONS

24. What is the pressure exerted by the point of a pencil held loosely in a vertical position? Make the necessary measurements to find an approximate answer.

25. What is the pressure on the bottom of your feet? To find out, trace your footprint (not your shoe) on a piece of paper and estimate its area. Consider your footprint as two rectangles, one for the heel and a larger one for the ball of the foot.

26. The sole of a snowshoe has a much larger area than that of a regular shoe. Why does a snowshoe make it easier to walk in deep snow?

27. The maximum force that human beings can exert with their jaws varies from person to person. Studies have shown that, on average, this force is 115 N. What is the greatest pressure that is exerted when biting on a hard carrot? (Hint: Bite very gently on the thick end of a carrot to estimate the initial area of contact between your teeth and the carrot.)

28. Suppose water is at the same level in two cylinders set up like those in Experiment 2.4. One cylinder has a base area of 2.0 cm^2 and the larger cylinder has a base area of 4.0 cm^2. If you remove 10 cm^3 from the larger cylinder, what will happen to the level of water in both cylinders?

29. Why are town water tanks built high up in the air or, if placed on the ground, located on hilltops?

30. The largest cruise ship built to date weighs 142,000 tons. This is about three times the weight of the Titanic! (1 ton = 2,000 lb)

    a. What is the buoyant force of the ocean on the ship?

    b. What is the weight of the seawater displaced by the ship?

    c. If the ship were able to sail into one of the fresh-water Great Lakes, what would be the buoyant force?

    d. How would the volume of water displaced in fresh water compare with that displaced in seawater? Explain your answer.

31. When a 490-N cedar log floats in a river, half the log is submerged.

    a. What is the buoyant force on the log?

    b. What is the weight of the displaced water?

    c. What is the weight per unit volume for cedar?

32. If you have played with a beach ball in a pool or the ocean, you may have noticed that it is quite difficult to submerge the ball completely. If a beach ball has a volume of 1.0×10^4 cm^3, what is the buoyant force on the ball when it is completely submerged?

33. A corked bottle contains just enough sand so that when it is submerged in 30 cm of water, it stays at that depth. What would happen to the bottle if it were submerged in 2 m of water?

34. Consider a closed plastic balloon filled with helium. When the balloon is released on a mountain 2,000 m above sea level, it just floats above the ground. What would happen to the balloon if it were released at sea level? (Plastic balloons stretch very little.)

THEMES FOR SHORT ESSAYS

1. On the moon, a person weighs only 1/6 as much as on Earth but may not lose any muscle strength. Write a description of a basketball game played without spacesuits inside a closed building on the moon.

2. Consider this familiar lament: "There's so much pressure on me to pass the test!" Discuss the relationship between the way the word "pressure" is used in this sentence and how it is used in this chapter.

Chapter 3
Forces Acting in Different Directions

3.1 Representing Forces

Suppose that you come across a beautiful waterfall while hiking. You want to tell your friend how to find it. Which are better instructions for how to reach the waterfall?

(1) "From the large oak tree, walk two kilometers and then walk half a kilometer."

(2) "From the large oak tree, walk two kilometers north and then walk half a kilometer northeast."

As you can see, some measurements require a statement of direction in addition to the size of the measurement. Since your friend is able to walk in any direction, you must specify the directions to make the instructions useful.

Force is another quantity that involves direction. Since forces can be exerted in any direction, we need to do more than just state its strength when we describe a force. We also need to state its direction. Such a quantity, described by both its size and its direction, is called a *vector*.

A vector can be represented by an arrow that is drawn to scale and points in the appropriate direction (Figure 3.1). In the case of a force, the direction of the vector indicates the direction of the force, and the length of the vector represents the strength of the force. In other words, the length of the vector is proportional to the force. When drawn to the same scale, a vector that is twice as long as another vector represents twice as much force, one three times as long represents three times the force, and so on.

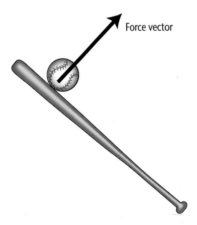

Force vector

Figure 3.1
When a bat hits a softball, it applies a force to the ball. The arrow is a vector that can be used to represent the force. The direction of the vector represents the direction of the force and the length of the vector represents the strength of the force.

Because the length of a force vector tells us about the strength of the force, we must be very careful to draw the length accurately. To draw a vector that is the proper length to represent a certain force, we must first choose a *scaling factor*. This scaling factor is the proportionality constant relating the length of the vector to the strength of the force. It states the length of a vector needed to represent each newton of force. You can choose any value for this scaling factor, but it will be helpful to choose a scaling factor that both is easy to work with and results in vectors that fit on your paper. For example, you might choose a scaling factor of 2 centimeters per newton (2 cm/N). Using this scaling factor, a 1-N force is represented by a vector that is 2 cm long, as shown in Figure 3.2(a).

Now suppose that you want to use the same 2 cm/N scaling factor to represent a force of 4 N (Figure 3.2(b)). A 4-N force has four times the strength of a 1-N force. Therefore, the vector drawn to represent the 4-N force is four times as long as the vector that represents the 1-N force, or 8 cm. Similarly, a 5-N force is represented by a vector five times as long (10 cm), and the vector for a 10-N force is ten times as long (20 cm).

In general, you can determine the proper length for the vector that represents any given force by using the following equation:

$$\text{Length} = \text{scaling factor} \cdot \text{force}$$

To check the length of your 4-N vector mathematically, you multiply the scaling factor times the force in newtons.

$$\text{Length} = (2 \text{ cm/N}) \times (4 \text{ N}) = 8 \text{ cm}$$

This is four times the length of the 1-N vector. If you use the same scaling factor to represent a force of one-half newton (Figure 3.2(c)), the length of the vector is

$$\text{Length} = (2 \text{ cm/N}) \times (0.5 \text{ N}) = 1 \text{ cm}$$

Suppose you want to work backward from the vector to determine the strength of the force. To do this, we rewrite our equation as:

$$\text{Force} = \text{new scaling factor} \cdot \text{length}$$

Your original scaling factor was 2 cm/N. This means that 1 centimeter represents 0.5 newton, giving us a new scaling factor of 0.5 N/cm.

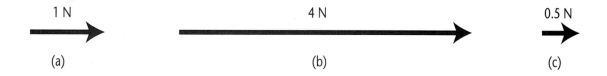

1 N

4 N

0.5 N

(a)

(b)

(c)

Figure 3.2
(a) Using a scaling factor of 2 cm/N, the vector that represents a 1-N force is 2 cm long. (b) Using the same scaling factor, the vector that represents a 4-N force is four times as long, or 8 cm. (c) By this scaling factor, a 0.5-N force is represented by a vector that is one-half the length of the 1-N arrow, or 1 cm.

How many newtons are represented by a vector 6 cm long? You can see that

$$\text{Force} = (0.5 \text{ N/cm}) \times (6 \text{ cm}) = 3 \text{ N}$$

For the given scaling factor, a 6-cm vector represents a force of 3 N.

1. Using a scaling factor of 5 cm/N, draw a vector to represent each of the following forces. (Hint: Since the length of a vector represents the size of a force, you must be careful not to make the vectors any longer when you add the arrowheads.)

   a. 2 N to the right

   b. 5 N toward the top of the page

   c. 1.5 N toward the top-left corner of the page

2. Using a scaling factor of 4 N/cm, describe each force represented in Figure A.

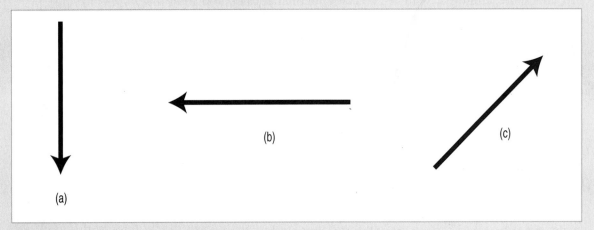

(a)

(b)

(c)

Figure A
For problem 2

3. Using a scaling factor of 0.2 cm/N, complete the following chart.

Size of force (N)	Length of vector (cm)
5	?
16	?
?	0.7
?	5.8

4. a. Using a scaling factor of 5 N/cm, complete the following chart.

Length of vector	Size of force	Direction
15 cm	?	upward
3.4 cm	?	west
?	25 N	north
?	18 N	downward

b. Draw each of the vectors specified in the chart.

5. How long should a vector be to represent a 14-N force if the scaling factor is 0.5 cm/N? How long should it be if the scaling factor is 0.2 cm/N?

EXPERIMENT
3.2 Combining Forces

In Chapter 1 you learned that if we exert a force on an object and it does not move, there must be an equal force in the opposite direction. We say that the forces on the object are *balanced*. To represent the two balanced forces, we can use two vectors of the same length pointing in opposite directions. We can now use this "balanced forces" idea to study how several forces combine when they act on an object. To begin, set up the apparatus as shown and explained in Figure 3.3.

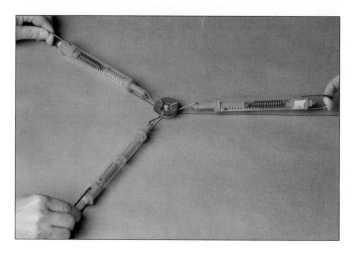

Figure 3.3
The force apparatus. On a large piece of paper, draw three line segments that meet at the center of the paper. Place a piece of wood or heavy cardboard under the center of the paper. Insert a pushpin through the paper and into the wood or cardboard at the point where the lines meet. Slip a large washer over the pushpin and hook three spring scales onto the washer. You will pull the spring scales along the lines you have drawn.

Label each line drawn on the large sheet of paper with the letter written on the spring scale that you will pull along that line. Then, with the help of your lab partner(s), pull the spring scales along the lines you have drawn. As you pull, be sure that the washer remains stationary over the pushpin without touching it.

- Why is it important that the washer and the spring scales not touch the pushpin as you take your force readings?

When the washer is steady over the pushpin, each lab partner should carefully read all three spring scales. Record these force readings next to each line on the paper.

- Are the forces on the washer balanced? How do you know?

To analyze the forces, you will use vectors. The lines you drew on the large sheet of paper provide the directions. But you still must choose a scaling factor and determine how long to draw the vector that represents each force. Remember to choose a scaling factor that both is easy to work with and allows you to draw vectors that fit on your paper.

Write the scaling factor that you have chosen on your paper. Then use this scaling factor to determine the appropriate lengths for the force vectors. After removing the pushpin, draw a vector along each line on your paper to represent the force measured in that direction. The starting point, or tail, of each vector should touch the point from which the pin was removed.

Before you can go any further with your analysis of the spring scale forces, you must understand how vectors combine. In Figure 3.4, the solid black vectors represent three forces acting on an object much like the washer you used in your experiment. The blue vector represents an imaginary force. This force has the same strength as force A but is in the opposite direction, so it balances force A. We also know that if all three forces (A, B, and C) are balanced, then the combination of forces B and C must balance force A.

Figure 3.4

If the three forces are balanced, the combination of forces B and C must balance force A. So the combination of forces B and C must be the same as the force represented by the blue vector. Notice that this vector is the diagonal of a parallelogram formed using the vectors for forces B and C as two of its sides. For that reason, this procedure for combining forces is known as the *parallelogram method*.

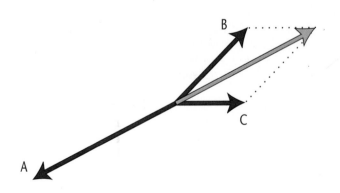

If force A is balanced by the combination of forces B and C, and the force represented by the blue vector also balances force A, then the force represented by the blue vector must be the same as the combination of forces B and C.

Notice another thing about Figure 3.4 — the blue vector is also the diagonal of a parallelogram formed from the vectors that represent forces B and C. This fact provides a way to graphically combine pairs of forces. The blue diagonal vector is the combination of the vectors representing forces B and C.

Now look again at the vectors you drew for the forces on the washer in your experiment. Draw a parallelogram using any two of the vectors. Then draw a new vector along the diagonal of the parallelogram. Repeat this procedure for each pair of original vectors on your paper. (You will end up drawing three parallelograms.)

- How does each vector formed along the diagonal of a parallelogram compare with the original vector that was *not* used to make that parallelogram?

- Suppose the combination of two forces acting on an object is the opposite of a third force acting on the object. What can be said about these three forces?

Whenever three forces are balanced, the combination of any two must balance the third. We can also say that if the combination of any two forces does not balance the third, then the three forces are not balanced.

- What can you say about the motion of an object that is initially at rest when three balanced forces are applied to it?

- What would you expect to happen to an object that is initially at rest when three *unbalanced* forces are applied to it?

6. **How do you draw the vectors to represent two forces that balance each other?**

7. **Suppose that you are given a vector diagram representing three forces applied to a soccer ball. When two of the vectors are combined, they form a new vector that is the opposite of the third original vector. What can you say about the forces on the soccer ball?**

8. **Four forces are applied to an object at rest. If the forces are balanced, what can you say about the combination of any two of the forces compared with the combination of the other two forces?**

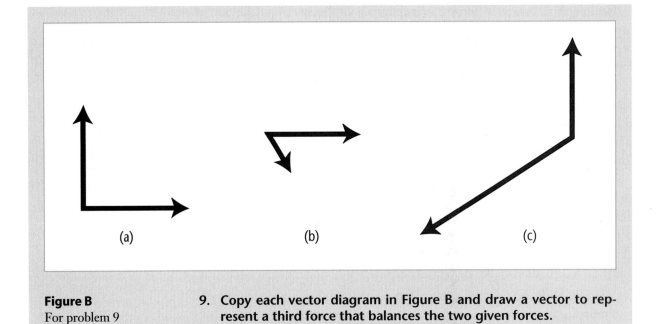

Figure B
For problem 9

9. **Copy each vector diagram in Figure B and draw a vector to represent a third force that balances the two given forces.**

3.3 The Net Force

Have you ever played soccer? If you have, you know that several people might kick a stationary ball at the same time. As a result of so many forces, the ball may not go where any of the players want it to go!

In Experiment 3.2, Combining Forces, you studied balanced forces. But what if the forces on an object at rest — like that soccer ball — do not balance? This question really has two parts: (1) What is the combined force on the ball? and (2) If the ball begins to move, in what direction will it move?

As you analyzed the forces in Experiment 3.2, Combining Forces, you used parallelograms to combine force vectors. This method has its limitations. It works quite nicely when combining two forces at an angle — that is, two forces that are not along the same line. But we need to devise a method for combining any number of forces, whether they are along the same line or at an angle. To do so, we must look at vectors, and the combination of vectors, in a slightly different way.

Since the important features of a vector are its length and its direction, all vectors that have the same length and direction are considered to be the same vector (Figure 3.5). Consequently, you can move a vector anywhere you want as long as (1) it stays the same length and (2) it points in the same direction (Figure 3.6). This second requirement means that the newly drawn vector must be parallel to the original vector.

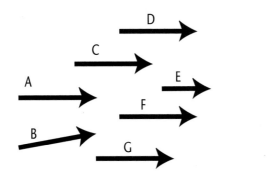

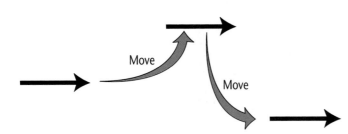

Figure 3.5
Vectors A, C, D, F, and G all are the same length and point in the same direction. They can all be thought of as the same vector. Vector B is the same length as A, C, D, F, and G, but it does not point in the same direction, so it is not the same vector. Vector E is a different length, so it is also a different vector.

Figure 3.6
We can move a vector to any position as long as it stays the same length and still points in the same direction. This means that the moved vector will be parallel to the original vector.

Figure 3.7(a) shows how to combine the vectors for two forces using the parallelogram method that you learned in Section 3.2. In a parallelogram, opposite sides are parallel and the same length. Because of this, we can imagine moving one of the vectors so that it forms a head-to-tail chain with the other vector (Figure 3.7(b)). A new vector, called the *resultant*, then represents the combination of the two original vectors. The resultant extends from the tail of the first vector in the chain to the head of the last vector in the chain, as shown in Figure 3.7(b).

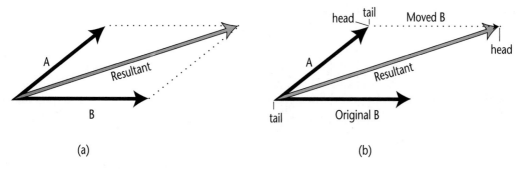

(a) (b)

Figure 3.7
(a) Combining two forces using the parallelogram method. The blue vector represents the combination of the two forces. (b) Notice that we get the same result by moving one of the vectors to create a head-to-tail chain and drawing the resultant from the tail of the first vector in the chain to the head of the last vector in the chain.

This head-to-tail method of combining vectors lets us combine any number of vectors whether they are along a line or at an angle. Figure 3.8 shows how the vectors that represent forces along a line are combined. Figure 3.9 shows the method for combining three forces at an angle. In each case, the resultant, shown in blue, represents the combined force.

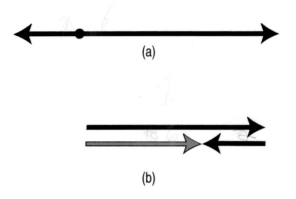

Figure 3.8
If two forces are applied to an object along a line (a), we can combine them by forming a head-to-tail chain with their vectors (b). The blue resultant is drawn from the tail of the first vector in the chain to the head of the last vector in the chain.

The combined force on an object is known as the *net force*. The net force on the soccer ball mentioned at the beginning of this section is a single force equivalent to the combination of all the forces applied to the ball.

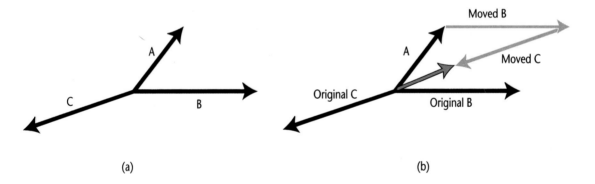

Figure 3.9
Any number of vectors can be combined using the head-to-tail method. (a) The vectors representing three forces. (b) The blue resultant, drawn from the tail of the first vector to the head of the last vector in the chain, represents the combination of the three original vectors.

Now we can consider the second part of the question posed at the beginning of this section. If the soccer ball begins to move, in what direction will it move? From experience you know what happens when you push hard enough on an object that is initially at rest. The object will begin to move in the direction of the push. When more than one force is applied to the object, the result is similar. If an object at rest begins to move, it will move in the direction of the net force on it. Therefore the stationary soccer ball kicked by three people at the same time will begin to move in the direction of the net force acting on it.

10. Each of the following sets of forces is applied to a ball. Determine the net force in each case.

a. 6 N upward, 4 N downward

b. 8 N to the right, 12 N to the left

c. 10 N downward, 5 N upward, 16 N upward, 7 N downward

11. If the scaling factor for Figure 3.7 is 2 N/cm, what is the net force? (Recall that you must specify both the strength of the force and its direction.)

12. In Figure C, the sets of vectors represent groups of forces applied to objects that are initially at rest. Determine the net force on each object.

Figure C
For problem 12

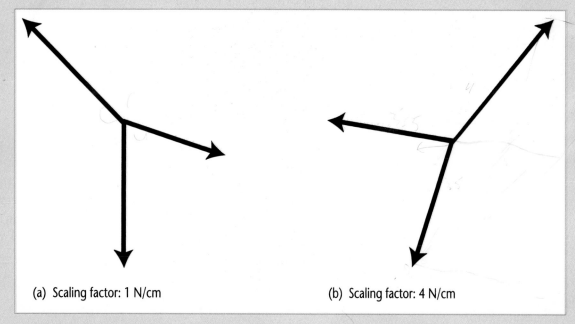

(a) Scaling factor: 1 N/cm   (b) Scaling factor: 4 N/cm

13. In Figure D, the four vectors represent four forces on a basketball that is initially at rest. Will the ball remain at rest or will it begin to move?

Figure D
For problem 13

3.4 Forces and Their Components

In the preceding section, you saw that forces acting on the same object combine to produce a net force. For example, in Figure 3.10, the combination of forces 1 and 2 is the same as force 3. Looking at this another way, we can think of a single force as being made up of two forces. As an example, we can think of force 3 as having been broken up into forces 1 and 2. Forces 1 and 2 are *components* of force 3.

Figure 3.10
Force 3 is the combination of forces 1 and 2, but it can also be thought of as being broken up into forces 1 and 2.

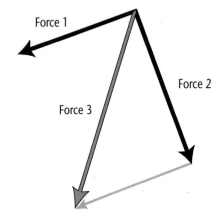

It is often useful to think of a single force as being made up of two perpendicular components. Suppose you want to find the components of a given force along two perpendicular directions (Figure 3.11). Draw a vector that represents the original force. Then draw lines in the two perpendicular directions (Figure 3.11(a)). By drawing two more lines, you can complete a rectangle that has the force as its diagonal (Figure 3.11 (b)). The sides of the rectangle represent the components of the force along the two perpendicular directions (Figure 3.11 (c)).

Figure 3.11
(a) The blue vector represents a force. The dotted lines specify two directions perpendicular to each other. (b) Completing the rectangle determines the size of the components in the two perpendicular directions. (c) The black vectors represent the perpendicular components of the original force.

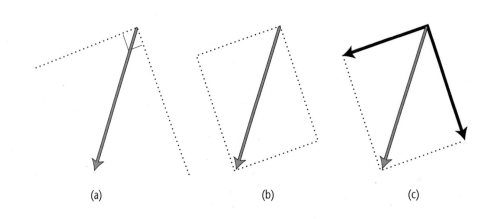

(a)   (b)   (c)

Consider the following situation. The crew of a delivery truck has to unload a heavy cart. The crew will not lower the cart straight down, but will let it roll very slowly down a ramp. The motion of the cart is controlled with a rope parallel to the ramp (Figure 3.12). With what force must a person pull on the rope?

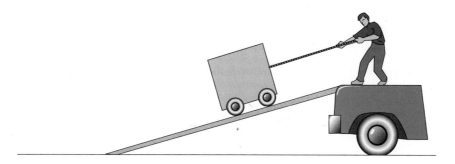

Figure 3.12
A heavy cart being slowly lowered from a truck. A rope held by a person on the truck controls the motion of the cart.

To answer this question, we can think of the weight of the cart as a combination of two components, one parallel to the ramp and one perpendicular to the ramp (Figure 3.13). When the cart is at rest, the net force on it is zero. That is, there must be two additional forces acting on the cart—one balancing force 1 and one balancing force 2. Force 1 is balanced by the pull of the rope. Force 2 is balanced by the ramp. We assume that the cart has low-friction wheels, so we can neglect friction. By drawing vectors to represent the forces, we can find the force with which the rope must be pulled.

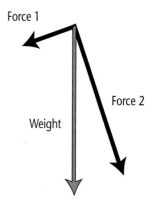

Force 1

Force 2

Weight

Figure 3.13
Looking at the weight of the cart as a combination of two forces, one parallel to and one perpendicular to the ramp.

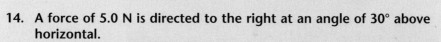

14. A force of 5.0 N is directed to the right at an angle of 30° above horizontal.

   a. Draw a vector representing this force using a scaling factor of 1 cm/N.

   b. What are the horizontal and vertical components of this force?

15. Suppose Figure 3.13 is a scale drawing of Figure 3.12 with a scaling factor of 1 cm/100 N.

   a. What is the weight of the cart?

   b. What is the force exerted by the rope?

16. Suppose that the cart in Figure 3.12 is replaced by a box of equal weight. Now friction cannot be neglected.

   a. What is the direction of the frictional force?

   b. How will the frictional force affect the force with which the rope has to be pulled when the box slides down very slowly?

EXPERIMENT
3.5 Forces Acting on Moving Bodies

You have now worked with balanced and unbalanced forces on objects that are initially at rest. What will happen if you apply a force to an object that is already moving? Will the object move in the direction of the force or will it move in some different direction? In this experiment, you will use an air puck, such as the one shown in Figure 3.14, to study these and other questions.

Figure 3.14

The assembled air-puck apparatus.

First clean the surface you will be using, since dirt and grit on the surface may interfere with the motion of the air puck. Then assemble the balloon and stopper apparatus as instructed in the caption for Figure 3.15.

Hold the mouth of the balloon over the large end of the stopper and blow into the coffee stirrer. When the balloon is inflated, remove the coffee stirrer. Then, using a slight twisting motion, press the small end of the

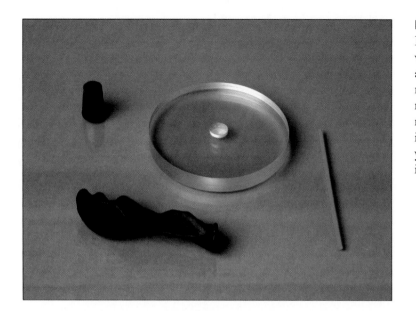

Figure 3.15
Fit the balloon over the wide end of the stopper and insert the coffee stirrer into the hole in the narrower end. To minimize the risk of spreading disease, always use your own coffee stirrer to inflate the balloon.

stopper into the hole on the top of the air puck. Position the air puck on the cleaned surface. By pressing down slightly on the air puck, you can prevent air from escaping until you are ready to make observations.

If the table you are using is not level, the air puck will drift "downhill." Use this direction as the initial direction of motion as you study the effects of forces on the air puck. The forces you apply will be provided by short puffs of air directed at the puck through a straw, as shown in Figure 3.16.

Figure 3.16
(a) Blow through a straw using short puffs directed at the air puck's base. (b) Be sure to point the straw at the center of the base, not off-center.

Keep each puff as short as possible, and always use the same strength puff at the same distance from the air puck.

Release the air puck so that it moves slowly in the downhill direction. (If your table is level, have your lab partner tap the air puck lightly with one finger so that the puck moves slowly in a straight line.) Then blow through a straw with short puffs and watch what happens to the air puck.

- If the puck is moving slowly toward you, what happens when you apply a force to it with a single short puff of air in the opposite direction?
- What happens when you apply a force to the puck with a single short puff in the same direction that it is already moving?

Now try applying short puffs at an angle to the direction of motion.

- How does the direction of the puck change after a puff at an angle?
- Does the puck move in the direction of the puff?

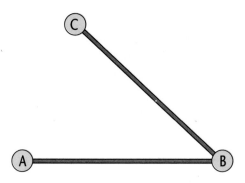

Figure 3.17
In what direction should you apply a single short puff through the straw to make the air puck follow the path shown?

Draw a simple path consisting of two line segments (similar to the one shown in Figure 3.17). Trade papers with your lab partner. If you are using a glass or clear plastic surface, you may place the diagram under it, but remember to re-level the surface before continuing. If you are placing the air puck directly on a tabletop, without a piece of glass or plastic, you will not be able to place the paper on the table without interfering with the motion of the puck. Instead, mark the path on the table using a water-based transparency pen. (Be sure to clean your marks off the table when you have finished.)

Experiment until you can make the air puck follow the marked path with only one short puff through the straw.

- Describe the direction in which you applied the force to make the air puck follow the marked path.
- State what you have learned about the direction of the force needed to cause a given change in the direction of motion of an air puck.
- Does an object always move in the direction of the net force applied to it?

17. An air puck is given a small push to start it moving. There are no other pushes or pulls on it until it reaches point X. At point X, a single puff of air is directed toward the puck through a straw.

   a. Which way should the straw be pointed to cause the air puck to follow the path shown in Figure E?

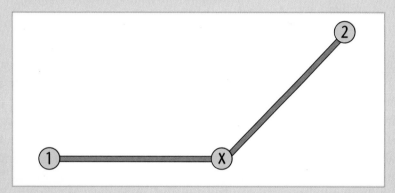

Figure E
For problem 17

   b. Which way should the straw be pointed to cause the air puck to follow the path shown in Figure F?

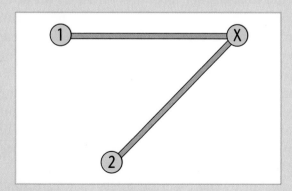

Figure F
For problem 17

3.6 Forces and Motion: A Summary

An air puck is hardly an object that you encounter in daily life. Nevertheless, the observations you made apply to a wide range of phenomena.

When the air puck is placed on a level, smooth table, it remains at rest. It does not move up or down, nor does it rest on the table; the upward force of the escaping air balances the downward force of gravity. There is no horizontal force acting on the puck. The net force on the puck is zero.

When the puck is given a short push, it starts moving. When the push stops, the net force is again zero. Although no net force is acting on the puck, it continues moving in a straight line at constant speed. These observations and many others similar to them lead to two conclusions:

(1) When the net force on an object at rest is zero, the object will remain at rest.

(2) When the net force on a moving object is zero, the object will continue moving at constant speed in a straight line.

The combination of these two conclusions is known as *Newton's first law*.

Newton's first law may seem strange to you at first. It appears to contradict your daily experience. You are used to applying a force just to keep things moving. For example, a moving shopping cart in the supermarket does not keep moving unless you keep pushing it. But in daily life, friction is always present. If the cart is moving in a straight line at a constant speed, the force you are applying simply balances the frictional force. The net force on the shopping cart is zero.

When you exerted a force on the puck in Experiment 3.5, you did so with a short burst of air. As you applied this force, the net force on the puck was not zero. But before and after the puff, the net force on the puck was almost zero because the force of friction was almost zero. This made it easier to observe the effect of your force on the motion of the puck.

You saw that when a nonzero net force acts on an object at rest, the object will start moving in the direction of the force. A force that acts in the direction that an object is already moving increases the speed of the object. A force acting in the opposite direction slows the object down and may even cause it to reverse its direction.

When the force on an object acts at an angle to the direction of motion of the object, the direction of motion changes. However, the new direction is not the direction of the force. The new direction is somewhere between the original direction of motion and the direction of the force. Soccer and hockey players are fully aware of this fact.

In the air puck experiment, you observed forces that were applied for very short periods of time. But the conclusions drawn from these observations hold for all forces, even those that act continuously, as when you push or pull on an object for a long period of time.

18. When a car is allowed to coast on a level road, it slows down and stops. When a car coasts up a hill, it stops and then starts rolling back down the hill. Identify the forces acting on the car in each of these situations.

FOR REVIEW, APPLICATIONS, AND EXTENSIONS

19. A vector that represents a force of 32 N is 8 cm long. What scaling factor was used to construct this vector?

20. Suppose a Boeing 777 is flying horizontally at a level altitude of 10,000 m. It is flying at a constant speed when the plane's engines provide a total forward force (thrust) of 640,000 N. The lift force holding the plane up is 2,500,000 N. Draw a vector that represents the combined effect of the thrust and the lift.

21. A child is holding the string attached to a helium-filled balloon. The string is stretched tightly and fully vertical.

    a. Is there a wind blowing on the balloon? Explain.

    b. What forces are acting on the balloon?

    c. Draw vectors to show the forces acting on the balloon.

    d. Is the buoyant force greater than, equal to, or less than the combined weight of the balloon and the string?

22. Form a team of three to do the following experiment. Two of you hold a rope about 3 m long as tight as you can. The third student pushes down on the rope near the middle.

    a. Why is it so difficult to resist the downward push?

    b. Copy Figure G. Assume that each student holding the rope applies a force of 200 N, as shown in Figure G. What is the strength of the downward force that the third student has to apply to hold the rope down? The scaling factor in Figure G is 50 N/cm.

200 N   200 N

Rope

Figure G
For problem 22

23. The air puck in Figure H is tied to a string and moving in a circle. If the string is cut when the puck is at point X, what will be the path of the puck?

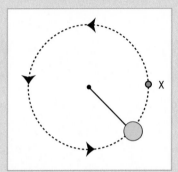

Figure H
For problem 23

THEMES FOR SHORT ESSAYS

1. In daily life, forces help us move, and their absence may be harmful. For example, a car may skid on an icy road. Describe a situation from your life when a missing force resulted in an unexpected motion.

2. Athletes need not know the laws of motion to use them in ice hockey, tennis, etc. But engineers need to understand these laws to launch a satellite. Tell why you think these situations are different.

Distance, Time, and Speed

Chapter 4
Distance, Time, and Speed

4.1 Introduction to Black Boxes

Today's telephone network is a very complex system. Very few people know everything about how it works. But almost everyone knows how to use it. People use telephones all the time to make local and long-distance calls. They know that telephones do what they are expected to do, so they have confidence in the system.

The same is true for digital watches. People use watches to read the time or to set alarms without knowing how they work. They trust digital watches because they can check them against established time standards.

Devices that we know how to use without knowing how they work are called *black boxes*. In the next experiment, you will use a motion detector to make a large number of distance measurements at short time intervals. The motion detector is a black box for you because you do not know how it works. To be sure that the motion detector does what it is expected to do, you will check its performance with a stopwatch and a meter stick.

Unlike a digital watch, a motion detector takes measurements whose results are not directly seen or heard. The data must first be processed by a computer or a graphing calculator (Figure 4.1). This is an advantage, though. The computer or calculator tabulates the measurements and plots a distance-time graph as the motion takes place. It does this much faster than you could do it with a stopwatch, meter stick, pencil, and graph paper. However, before you can trust these data and graphs, you must be sure that the motion detector gives the same readings as a clock and a meter stick.

Figure 4.1
The motion detector, computer interface, and computer. The screen shows a graph of distance versus time taken by this assembly.

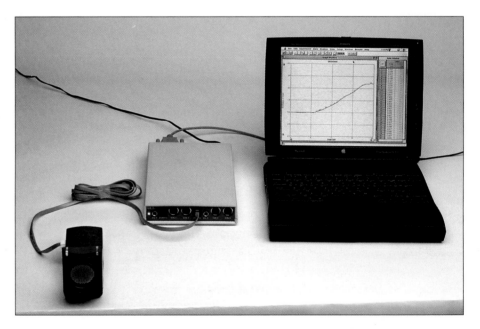

1. **Name one black box that you have in your home. Describe what it does and explain how you can verify that it does what it is expected to do.**

EXPERIMENT
4.2 The Motion Detector — Measuring Distance

The study of motion begins with the measurement of time and distance. To measure time, you can use a computer or a graphing calculator. These instruments have internal clocks. Their time measurements can be easily checked by a stopwatch or a classroom clock.

To measure distances, you can use a motion detector. But before you can trust those measurements, you must first convince yourself that the motion detector gives the same distances you would obtain with a meter stick. To do this, first create a distance line on the floor by using a meter stick to measure 0.00, 1.00, 2.00, and 3.00 meters from the edge of your lab table (Figure 4.2). You may want to use a strip of masking tape to mark these distances on the floor. (To make the tape easier to remove, fold over one end onto itself to create a tab that will not stick to the floor. You can then use this tab to remove the tape when you have finished the experiment.)

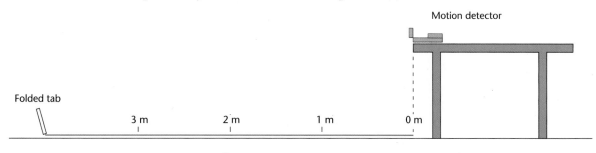

Motion detector

Folded tab

3 m   2 m   1 m   0 m

Tape

To see if the motion detector reads a distance correctly, set it on your lab table and point it horizontally in the direction of the distance line you have just created. Then stand straight and still with your heels at the 2-m mark, facing away from the detector. To prevent interference, be sure no other objects are within a meter on either side of the distance line on the floor. Have your lab partner operate the motion detector as directed by your teacher. The motion detector is taking readings when you hear it clicking. As the detector takes readings, the computer produces a graph that shows distance on the vertical axis and time on the horizontal axis.

Figure 4.2
The experimental setup.

- Why is it important to measure the distance to the 2-m mark exactly?
- Why is it important to stand still and straight at the 2-m mark?
- Why does the graph show a horizontal line?

If the horizontal line is not at 2.00 m, adjust the position of the motion detector by moving it forward or back until it reads 2.00 m when you stand at the 2-m mark. When you are sure that the motion detector reads 2.00 m, tape it to the table so that it does not move during the remainder of this checking procedure.

Next, you need to see if the motion detector produces divisions that are the correct size. In other words, will you still obtain a correct reading if you change your distance from the detector? To check this, stand now at the 3-m mark while your partner operates the motion detector.

- Does the motion detector read 3.00 m?
- How does the graph show that you increased your distance from the detector by 1.00 m?

Repeat the procedure as you stand 1.00 m from the detector.

- What distance does the motion detector read when you stand at the 1-m mark?
- Are there any differences between the detector readings and the meter-stick measurements?

To realize the sensitivity of the motion detector, you can observe how far an object must move in order for its motion to be detected. Stand at the 2-m mark while your partner operates the motion detector. During a 10-s interval that the detector is on, lean forward slightly and then back without moving your feet.

- Can you detect the leaning motion on the graph?
- Use the graph to estimate how far forward your partner leaned.
- How does your estimate compare with the data table on the right side of the graph?

To see if the clock in your computer measures time correctly, you can compare it with a watch. Perhaps you or one of your classmates has a watch with a second hand or, better yet, a stopwatch feature. Turn on the detector for a 10-s run. When the distance line is at 2.00 s, start the stopwatch and measure the time required for the line to reach the 10-s position on the graph.

- How do the two times compare?

If you have determined that the motion detector gives the same readings as a meter stick, you can trust it to make the distance measurements that are necessary to study motion.

2. **In Figure A, how far from the motion detector was the person standing?**

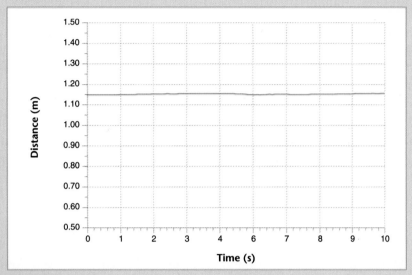

Figure A
For problem 2

3. **Describe what you think may have happened to produce the graph shown in Figure B.**

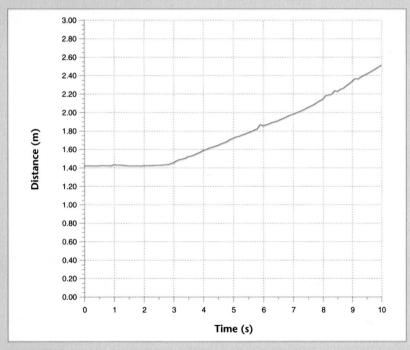

Figure B
For problem 3

EXPERIMENT
4.3 The Motion Detector — Motion Graphs

In Section 4.2, you saw that when you stand still in front of a motion detector, a horizontal line appears on the distance-time graph. This line is the result of many nearly identical distance measurements during the period of time in which the motion detector took readings.

Now, beginning at a distance of 1 m from the detector, walk away slowly using shuffling "baby steps." If you get a very rough, jagged graph, just try again. It only takes a few seconds! (Hint: To avoid such a "noisy" graph, you might try keeping your arms at your sides or in front of you, close to your body.)

• How can you tell which portion of the graph represents your motion as you were walking?

When you look at a motion graph, you need to concentrate on the portion of the graph that represents the motion you want to analyze. In this experiment, that portion is a straight line.

• What are the beginning and end times for the relevant portion of your motion graph?

Next, starting from the same 1-m mark, collect data as you walk away from the motion detector at a slightly faster pace. Then repeat this procedure once more while walking even faster.

• What are the beginning and end times for the relevant portions of your motion graphs?

• How can you tell from these graphs which motion was faster?

Now, beginning at a distance of 3 m from the detector, walk slowly toward the detector.

• Which portion of this graph shows your motion during this walk?

• How is this graph different from the graphs made when you were walking away from the motion detector?

> 4. Figure C shows the motion graph for a simulated "walk" in front of a motion detector.

a. Suppose you want to reproduce the motion shown by the graph. During which segment, B or D, must you move faster? How is this shown on the graph?

b. Look at segments D and F on the graph. Which segment represents motion away from the detector and which represents motion toward it? How is this difference shown on the graph?

c. Imagine that you are trying to tell someone how to move in front of the motion detector in order to reproduce this graph. What instructions should you give?

d. The graph shows several sharp angles where the different segments of the graph meet. Do you think it is possible to move so that you create such sharp angles? Why or why not?

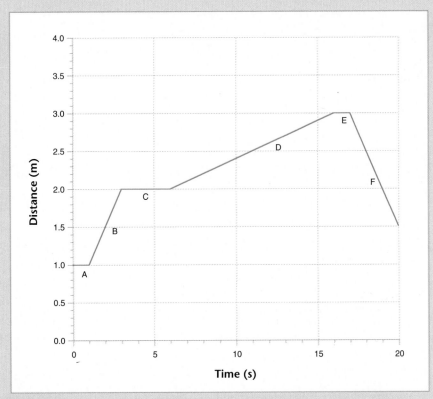

Figure C
For problem 4

5. Describe in detail how you should walk to make a graph similar in shape to the one shown in Figure D.

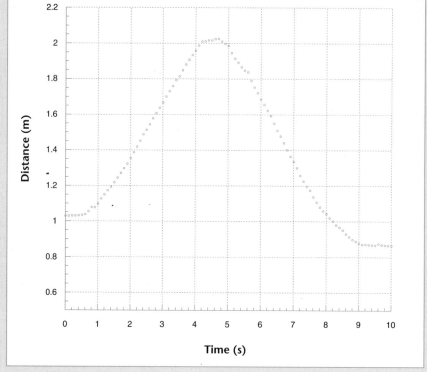

Figure D
For problem 5

4.4 Distance, Time, and Average Speed

By walking in front of a motion detector, you can produce a detailed description of the motion. When you hike on a trail or ride in a car, you can not provide such a detailed description. To get a general idea of how fast you have moved, divide the total distance you moved by the time it took to move that distance. This quotient is called the *average speed*:

$$\text{Average speed} = \frac{\text{total distance}}{\text{total time}}$$

The unit of average speed is the unit of distance divided by the unit of time. If the distance is measured in meters (m) and the time in seconds (s), then the average speed is expressed in meters per second (m/s). If the distance is measured in kilometers and the time in hours, the average speed is expressed in kilometers per hour (km/h).

Figure 4.3

A graph of distance versus time for a simulated trip by three cars. To make a graph of a real car trip, you could record the odometer reading every minute and then draw the graph from the time-and-distance table.

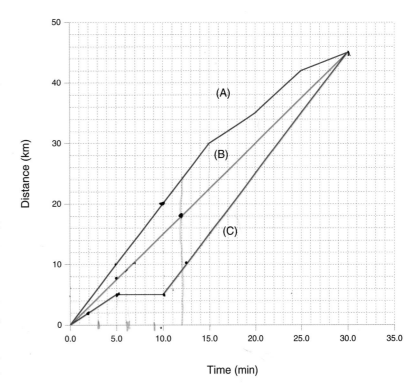

The distance traveled need not be along a straight line. If you walk around the block, you end up where you started. Nevertheless, your average speed equals the total distance you walked divided by the total time it took.

Average speed gives only a general idea of how fast we travel. It lacks any detail. This can be seen in Figure 4.3. All three cars started at the same time from the same place. They arrived at the same destination at the same time. Their average speed is the same because they covered the same distance in the same time. Yet, along the way, they were driven quite differently. Car A went faster than the other two cars during the first twenty minutes, then it slowed down. Car C stopped for five minutes early in the trip.

Average speed can be calculated over different segments of the trip. Table 4.1 lists the average speed for the three cars over 5-min intervals.

Table 4.1 Average Speeds of Cars in Figure 4.3 over 5-min Intervals

Time interval (min)	Car A (km/min)	Car B (km/min)	Car C (km/min)
0– 5	2.0	1.5	1.0
5–10	2.0	1.5	0.0
10–15	2.0	1.5	2.0
15–20	1.0	1.5	2.0
20–25	1.4	1.5	2.0
25–30	0.6	1.5	2.0

Generally, average speed depends on the time interval over which it is calculated. Table 4.2 lists the average speeds for the same three cars over different time intervals. Notice that only car B had the same average speed in all intervals. The average speeds of the other two cars are different in some time intervals. When the average speed of an object is the same during any time interval during the motion, the object is said to be moving at a *constant speed*. In Figure 4.3, only car B was moving at a constant speed all the time. On a graph of distance versus time, motion at a constant speed appears as a straight segment.

Table 4.2 Average Speeds of Cars in Figure 4.3 over 10-min Intervals

Time interval (min)	Car A (km/min)	Car B (km/min)	Car C (km/min)
0–10	2.0	1.5	0.5
5–15	2.0	1.5	1.0
10–20	1.5	1.5	2.0
15–25	1.2	1.5	2.0
20–30	1.0	1.5	2.0

6. Using Figure 4.X, verify the entries in rows 2 and 5 of Table 4.1 and the entries in rows 1 and 5 of Table 4.2.

7. Using Figure 4.3, divide the first 12 min shown on the graph into four 3-min intervals.

   a. What distance did car C cover in each 3-min interval?

   b. Calculate the average speed of car C during each 3-min interval.

8. The Boston Marathon is run over a course of 42.19 km. A well-conditioned runner can finish the race in 2.5 h. What is the average speed of such an athlete during the race?

9. A racecar driver drove a racecar one lap around a track in 41.55 s. The racetrack is 2.50 mi long. (You may want to use Appendix 3 Conversion of Units.)

   a. What is the car's average speed in miles per hour?

   b. What is its average speed in meters per second?

EXPERIMENT
4.5 Terminal Speed

When you ride a bicycle starting from rest, you very quickly reach a speed that you want to keep constant for most of the ride. For cars, trains, airplanes, and so on, this speed is called the *cruising speed*. Does a "cruising speed" occur only through human control, or does it also happen in nature?

To find out, you will investigate how a coffee filter falls to the floor after being released (Figure 4.4). Be sure to have the protective cover in place to avoid accidental damage to the detector. In deciding from which position to release the filter, take two considerations into account. You want the filter to fall over the longest possible distance, but you also want the detector to "see" only the filter, not you as well.

- How should you hold the filter to achieve a reasonable compromise between these two considerations?

- After releasing the filter, why should you keep your arms up until the filter reaches the floor?

With your partner, try a run or two to see how the graphs look. The motion detector only registers the location of objects at a distance greater than 40–50 cm.

- How does the height-versus-time graph show that the motion detector loses sight of the filter when it is closer than 40–50 cm?

Change the time axis so that only the part where the motion of the filter has been recorded is visible.

- In what time interval did the coffee filter fall at constant speed? How did you reach your conclusion?

The constant speed reached by a falling body is called the *terminal speed*.

- What was the terminal speed of the coffee filter?

*   *   *

Figure 4.4
A coffee filter shortly after being released. Note the position of the student's arms and the position of the motion detector.

There are only two forces acting on the filter once you release it — the gravitational force and the friction with the air. When you released the filter, the force of friction exerted by the air must have been less than the gravitational force. If the force of friction had been as strong as the gravitational force, the filter would have stayed suspended in midair. It did not. The filter fell and reached a terminal speed in a very short time. This tells us that the frictional force exerted by the air increased as the speed of the filter increased until it balanced the gravitational force. Generally, the force of friction increases with speed as a solid object moves through a gas or a liquid.

10. Figure E is a graph of height versus time for a falling coffee filter. The motion detector registered the height every tenth of a second. From the graph, find the time at which the filter reached terminal speed.

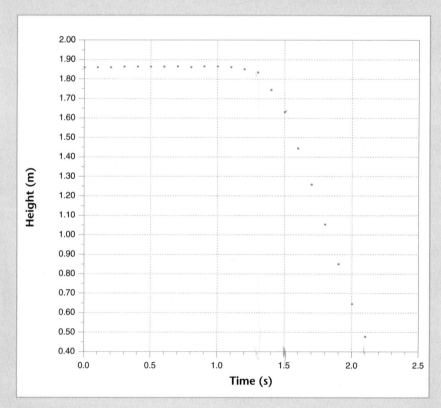

Figure E
For problem 10

4.6 Working with Distance, Time, and Constant Speed

Why was it worthwhile to study the motion of a falling coffee filter? The answer is that most motions in daily life begin at rest, reach approximately a constant speed, and end at rest. It takes you just one or at most two steps before you are walking at constant speed. It takes no longer to come to rest. In a fifteen-minute walk, you make so many steps at constant speed that the few steps at the beginning and at the end can be ignored.

In normal driving, it took 12 s for a car initially at rest to reach a speed of 22 m/s (50 mi/h). The distance covered during this 12-s interval was about 160 m. These are very short times and distances compared with the duration of a trip and the distance covered. The same is true for the time and distance it takes to slow down at the end of a trip. Therefore, we can think of a trip on a highway as consisting of periods of rest and motion at constant speed.

Speed can be measured directly, unlike average speed. The speedometer on a car does just that. Thus, when any two of the three quantities — time, distance, and constant speed — are known, the third quantity can be calculated. For constant speed, from

$$\text{Speed} = \frac{\text{distance}}{\text{time}}$$

it follows that

$$\text{Distance} = \text{speed} \cdot \text{time}.$$

and

$$\text{Time} = \frac{\text{distance}}{\text{speed}}$$

The last equation is especially useful for calculating how long a trip may take. For example, a 240-mi trip on a highway at 60 mi/h, will require (240 mi)/(60 mi/h) = 4.0 h travel time. On such a trip, there may be one or two rest stops totaling 0.5 h. This increases the total travel time to 4.5 h. The average speed will be (240 mi)/(4.5 h) = 53 mi/h.

11. A safe driver plans on driving at a constant speed of 60 mi/h. He also plans to make three 10-min rest stops and take one hour for lunch during his 10 h on the road. How far will he travel during the day?

12. A driver set her cruise control at 70 mi/h during the 208-mi trip from Memphis to Nashville. She left Memphis at 1:13 P.M. and arrived in Nashville at 4:27 P.M. She took one rest break.

a. How long did her trip take?

b. What was her average speed?

c. Approximately how long was her rest break?

13. What is the minimum time a driver could travel at constant speed between two toll stops 26 mi apart and not violate the 65 mi/h speed limit?

14. Current large passenger airplanes can transport over 400 passengers at a maximum speed of 560 mi/h with negligible wind. The range of such an airplane is 5,950 mi.

a. What is the maximum speed of this airplane in meters per second?

b. What is the minimum time in which this airplane could fly its range?

FOR REVIEW, APPLICATIONS, AND EXTENSIONS

15. The speedometer in your family's car may be a black box for you. Suggest a way to check that it works properly.

16. The graph shown in Figure F on the next page represents the motions of three cars — car 1, car 2, and car 3.

a. Using the graphs, calculate the average speed of each car over the entire interval shown.

b. Which, if any, of the three cars stopped during the interval?

c. Which car has the highest average speed between 20 min and 30 min?

17. If you flew at a constant speed of 800 km/h between Boston and San Francisco, how long would the trip take?

18. Using a map, select two towns that are at least 150 mi apart. Determine the distance between them. If you could maintain a constant speed of 55 mi/h, how long would it take to travel between the towns?

19. Choose another two towns that are more than 200 mi apart. If you traveled at a constant speed of 50 mi/h but stopped an hour for lunch somewhere along the way, how long would it take to make the trip?

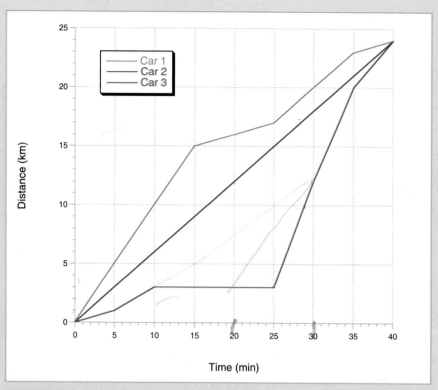

Figure F
For problem 16

20. The table shows the racing distance and time for men's world-record holders.

World Records for Men's Running (2001)

Distance (m)	Time (min)	(s)	Runner
800	1	41.11	Wilson Kipketer (Denmark)
1,500	3	26.00	Hicham El Guerrouj (Morocco)
3,000	7	20.67	Daniel Komen (Kenya)
5,000	12	39.36	Haile Gebreselasie (Ethiopia)
10,000	26	22.75	Haile Gebreselasie (Ethiopia)

a. What is the average speed of each runner during his race in meters per second? Graph average speed versus distance.

b. What conclusions can you draw from the overall shape of the graph?

21. The world records for women's freestyle swimming are shown in the table.

World Records for Women's Freestyle Swimming (2001)

Distance (m)	Time (min)	(s)	Swimmer
50		24.16	Inge de Bruijn (The Netherlands)
100		53.77	Inge de Bruijn (The Netherlands)
200	1	56.78	Franziska van Almsick (Germany)
400	4	3.85	Janet Evans (U.S.)
800	8	16.22	Janet Evans (U.S.)
1,500	15	52.10	Janet Evans (U.S.)

a. What time do you predict for Janet Evans in the 1,000-m freestyle race? (Hint: Draw a distance-time graph.)

b. What time would you predict for Inge de Bruijn in the 400-m race?

c. Are you equally sure of your two predictions? Why?

THEME FOR A SHORT ESSAY

Many birds of prey, such as falcons, are able to hunt more efficiently by increasing their terminal speed through the air. Some other birds and other types of animals, such as flying squirrels, are able to decrease their terminal speed. Using the Internet or other sources, choose one such animal and describe how it makes its adjustment.

Chapter 5
Waves

5.1 Sound: Something Else that Moves

Think about what happens when you walk from one classroom to another. At the end of this walk, you are no longer in the first classroom but in the second. If you bump into another student on the way, both your motions will be affected. Both of you may stop or, at least, change direction. What occurs during a walk like this is typical of the motion of all material objects — not just people. So far, every motion that we have considered has been the motion of some object. But are objects the only things that move?

Consider sound. Whether a sound comes from a violin, a drum, or a tuning fork, that sound is produced by something that vibrates. A violin string, a drumhead, and the arms of a tuning fork all vibrate and push against the air around them. Does the air that is being pushed move along with the sound as it travels to your ear?

Air is invisible, but we can use smoke to make its movements visible. Suppose there is a puff of smoke in front of the speakers on a radio. Turning on the radio does not affect the motion of the smoke. Sound moves through the air, but the air itself does not move along with it.

Another difference between the motion of sound and the motion of objects is illustrated by what happens when two moving objects or two sounds come together. When moving objects collide, they change each other's motion. Now think about people sitting around a table while several conversations go on at the same time. Although the conversations may "cross," the sounds produced by the speakers do not speed up, slow down, or become distorted. Apparently, sound passes through sound unaffected.

There are other differences between the movement of objects and the movement of sound. Friction causes moving objects to slow down and eventually to stop. Yet, with measurements of the type that you will perform shortly, we find that sound weakens with distance from the source but does not slow down.

Sound moves through water as well as through air. By putting some dye in water, we can show that the water does not move along with the sound passing through it, just as air does not move with the sound passing through it. However, a doorbell placed inside a container from which all the air has been removed produces no sound. Sound needs a substance through which to move. The kind of motion exemplified by sound is called *wave* motion. Waves move very differently from material objects.

1. **During one of the Apollo moon missions, an astronaut hit a golf ball on the lunar surface. Could another astronaut standing nearby hear the sound of the golf club's impact on the ball? Explain.**

5.2 Visualizing Sound: Longitudinal Waves

Even when we place smoke in air or dye in water, we cannot see what happens when sound moves through each substance. A coil spring can help us to visualize the motion of sound.

When the end of a stretched coil spring is rapidly pushed inward and then pulled back, the distance between adjacent coils changes. While the end of the spring is pushed inward, the distance between adjacent coils decreases. The spring is now compressed. When the end of the spring is pulled back, the distance between adjacent coils increases, so that it is greater than when the spring was at rest. The spring is now decompressed. The result of two such motions is shown in Figure 5.1.

Figure 5.1

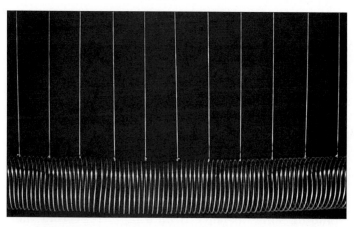

(a) A coil spring at rest. The spring is supported by many long threads.

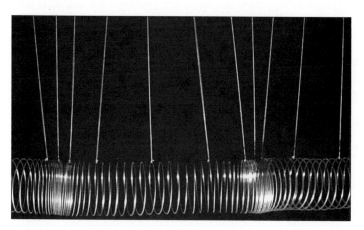

(b) The same section of the spring as two short waves pass through it. Note the alternating regions of compression and decompression.

The pattern you see in Figure 5.1(b) does not stay where it was created: it travels the length of the spring. The motion of such a pattern is a wave motion. The motion of a few coils of the spring is back and forth along the same line as the motion of the wave. A wave created by such motion is called a *longitudinal wave*. After a wave has passed a certain region of the spring, the spring returns to its original shape.

When a wave is generated at some point on the spring other than one of the ends, the wave moves outward from that point in both directions along the spring and may be reflected at the ends of the spring. As it travels, the wave dies out rapidly, just as sound fades, or dies out, with distance. However, the wave does not slow down.

Air behaves much like a coil spring extending in all directions. When air is pushed back and forth rapidly, longitudinal waves are created. Under certain conditions, we perceive these waves as sound.

> **2. Describe how a tuning fork creates a sound that reaches the ear.**

EXPERIMENT
5.3 The Speed of Sound

Suppose you are standing some distance from a large reflecting wall. When you make a sharp sound by clapping your hands or banging two pieces of wood together, you hear the original sound followed by an echo. In this experiment, you will measure the distance traveled by a sound to a wall and back, as well as the time it takes the sound to make that roundtrip. These measurements can then be used to calculate the speed at which sound travels.

Your teacher will direct you to a large wall that will produce a clear echo when you stand 40 to 60 m in front of it.

- Why is it important that there be no other reflecting surface nearby?
- What is the distance between you and the wall?
- What distance does the sound travel between the time you hear the clap and the time you hear the echo?

It is your partner's task to use a stopwatch to measure the time interval between the clap and its echo. Because the roundtrip time interval for one echo is so short, the error in this measurement will be relatively large. To increase the accuracy of the time measurement, you can do the following. Clap the blocks of wood together at regular time intervals and adjust the rate of clapping until each clap coincides with the echo of the preceding one. You may have to practice to achieve the rhythm that synchronizes the echo with the clap. When you have achieved the proper rhythm, you will not hear the echoes. Continue clapping until your partner measures the total time for 20 or 30 roundtrips.

- When you count the number of roundtrips that the sound makes, why should you count the first clap as "zero"?
- What is the time for one roundtrip?
- What did you calculate for the speed of sound in air?
- How could you extend the experiment to determine whether sound slows down as it moves farther away from its source?

If the conditions at your school permit, carry out your plan.

- What did you find?

---

3. When a group of students performed Experiment 5.3, The Speed of Sound, they stood a distance of 45.0 m from the wall. They found a total time of 5.10 s when they counted 20 roundtrips. What was the speed of sound from their experiment?

4. If the group of students in question 3 counted the first clap as "one" instead of "zero," would the error increase or decrease the value found for the speed of sound compared with the value found using the correct procedure?

5.4 Waves in Gases and Liquids

The kind of spring you use affects the speed of waves moving through it. Even the amount of stretch of the same spring affects the speed of the waves. Similarly, the speed of sound in a gas depends on the kind of gas through which it passes. In helium, for example, sound travels faster than it does in air. Even for the same gas, the speed of sound depends on the conditions. For example, temperature and humidity affect the speed of sound in air. However, the speed of sound does not depend on atmospheric pressure.

Sound waves also move through liquids. You may have noticed this if you have ever swum underwater. The speed of sound in a liquid depends on the kind of liquid used and its temperature. (See Table 5.1.)

Table 5.1 Speed of Sound in Some Gases and Liquids

Substance	Temperature (°C)	Speed of sound (m/s)
Helium	0	965
Air, dry	−10	326
Air, dry	10	337
Air, dry	30	349
Water, fresh	10	1,447
Water, fresh	20	1,482
Water, sea	10	1,490
Water, sea	20	1,522
Mercury	25	1,451

5. Suppose some students on a mountain field-trip find a gorge that returns a very clear echo when they shout. Several trials show that the sound returns in 3.00 s. The temperature is 10°C. How far away is the reflecting surface?

6. Suppose two people are swimming close to each other in a lake. One of the swimmers has her head underwater when a motorboat 200 m away starts its engine. Which swimmer will hear the engine first?

7. Compared with its speed in air, how many times as great is the speed of sound in water?

8. Suppose that during a thunderstorm you hear a clap of thunder 4.0 s after you see a bolt of lightning. The temperature is 20°C. How far away from you is the lightning? (*Note:* You may neglect the time it takes the light from the flash to reach you.)

EXPERIMENT
5.5 Transverse Waves on a Coil Spring

In this experiment, you will study a different kind of wave motion.

Part A

With the help of a lab partner, stretch a coil spring across the floor until it extends at least 5 to 6 m. Be sure the spring is straight. With your partner tightly holding one end of the spring, quickly and sharply move the other end of the spring sideways (perpendicular to the line of the spring) and then back to its starting point. Repeat this procedure with your partner moving the spring while you hold your end in place.

As in the case of a longitudinal wave, a pattern moves along the entire length of the spring. Figure 5.2 shows a series of photographs in which a ribbon is shown attached to one coil of the spring. The marked coil moved perpendicular to the line of the spring, just as your hand does when you generate a wave in this experiment. A wave in which the motion of the spring is perpendicular to the motion of the wave is called a *transverse wave*. Transverse waves are much easier to observe in detail than longitudinal waves.

You and your partner can each send a wave from the same or opposite ends of the spring at the same time. First send a large wave while your partner sends a smaller wave on the same side of the spring. Repeat with large and small waves sent on opposite sides of the spring. Watch closely as the waves move along the entire spring.

- How does the behavior of the waves compare with the behavior of two objects, such as basketballs, when they meet?

Part B

The waves that you generated did not continue moving indefinitely. They die out just as sound dies out. But do they travel at a constant speed as sound does, or do they slow down like a ball coming to rest because of friction?

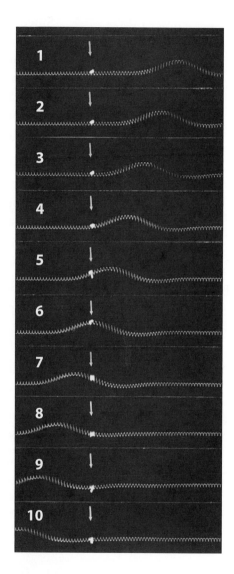

Figure 5.2
Photographs of a wave on a coil spring taken with a movie camera. A ribbon is tied to one coil of the spring. The ribbon moves up and down while the wave moves from right to left.

To find out, place a tape marker about 1 m from each end of a stretched coil spring. Enlist the help of four more classmates to time a wave traveling between these markers. Have two of the timers take a position behind you, and position the other two behind your lab partner. When the timers are ready, generate a wave and have the timers measure how long it takes the wave to move from one tape marker to the other.

Since the wave extends over some length, you must decide at which point to start and stop the stopwatch. You may find the midpoint of the wave to be the most convenient point for this purpose.

Repeat this measurement several times and average the results. Be sure to keep the spring stretched to the same length, while also keeping the same number of coils in your hand.

- What was the speed of the wave as it traveled along the spring?

To study what happens over a longer distance, generate a wave that travels all the way down the spring, reflects from the fixed end, and returns to its starting point. (If the wave dies out before making it all the way back, generate a larger wave.)

- What is the ratio of the speed of the reflected wave to the speed of the original wave? What does this ratio tell you about the speed of the wave?

Now try stretching the spring by a different amount. This can be done either by stretching the spring to a different length or by gathering up a different number of coils in your hand.

- Does the stretch of the spring play a part in determining the speed of the wave?

9. In which direction is the white ribbon moving in the fifth, sixth, seventh, and eighth frames of Figure 5.2?

10. Because of friction, objects such as bicycles require a continuous force to keep them moving. In the absence of this force, friction slows them down and causes them to stop. After you generate a wave on a coil spring, must you continue to apply a force to keep the wave moving at a constant speed? Explain.

11. The frames in Figure 5.2 were taken at regular time intervals with a movie camera in a fixed position.

    a. Did the wave slow down? How can you tell?

    b. Did the height of the wave diminish? How can you tell?

5.6 Waves in Solids

Both longitudinal and transverse waves on a coil spring move only along the spring itself. That is, they move in one dimension. Nevertheless, the coil spring serves as a good way to visualize waves in solids because solids carry both longitudinal and transverse waves. Suppose that you tap on a large block of iron. Both a longitudinal and a transverse wave will move through the block in all directions. Each wave has its own speed. The speeds of the two kinds of waves in some solids are given in Table 5.2. Notice that for each material, the speed of the longitudinal wave is greater than the speed of the transverse wave.

Table 5.2 Speeds of Longitudinal and Transverse Waves in Some Solids

Substance	Speed of longitudinal wave (m/s)	Speed of transverse wave (m/s)
Aluminum	6,420	3,040
Copper	4,760	2,110
Iron	5,950	3,240
Glass	5,640	3,280
Rock	6,000–13,000	3,500–7,500

The precise values for the speeds of the waves in the metals and in glass depend on the way the materials were prepared and on the amount and nature of any impurities. Thus, the values in the table are approximate, and the zeros serve only as placeholders. Since there are many kinds of rocks for which the speeds are known, the table provides a range of speeds for these substances.

The fact that only solids can carry both longitudinal and transverse waves is applied very effectively in locating the origin of an earthquake. The principle behind this application is as follows.

At the origin of an earthquake, the earth shakes vigorously, sending out both longitudinal and transverse waves in all directions. The two waves start out at the same time. However, an observer who is at some other location and is equipped with a clock and a detector does not know when the wave started. The detector registers only the time at which the wave

Figure 5.3

Distance-time graphs for longitudinal and transverse waves drawn on the same axes.

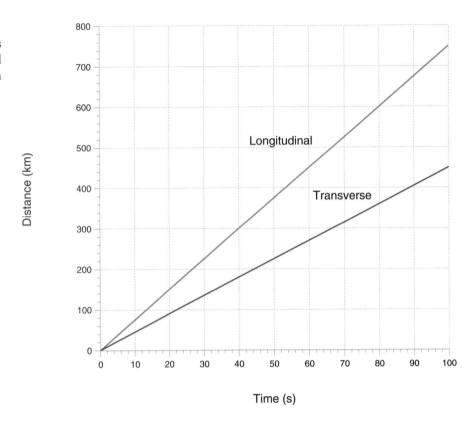

reached it. Since the travel time of the wave is not known, we cannot use the relation

$$\text{Distance} = \text{speed} \cdot \text{time}$$

to calculate the distance. How can this difficulty be overcome?

Figure 5.3 is a distance-versus-time graph for longitudinal waves (red line) and transverse waves (blue line) starting at the same time from the same location (time = 0 and distance = 0). As time passes, the longitudinal wave will be farther and farther ahead of the transverse wave. For example, after 40 s the longitudinal wave will have traveled 300 km, while the transverse wave will have traveled only 180 km.

How long does it take the transverse wave to travel 300 km? From Figure 5.3, you can see that the transverse wave will reach 300 km after 67 s. This is 27 s (67 s – 40 s) after the arrival of the longitudinal wave. The detector records this time difference.

Suppose that a transverse wave from an earthquake arrived at a detector 25 s after the longitudinal wave. How far from the origin of the earthquake was the detector? To answer this question, draw a segment that represents 25 s on the time axis. Then move the segment you have drawn parallel to the time axis on the graph until the horizontal distance between the red

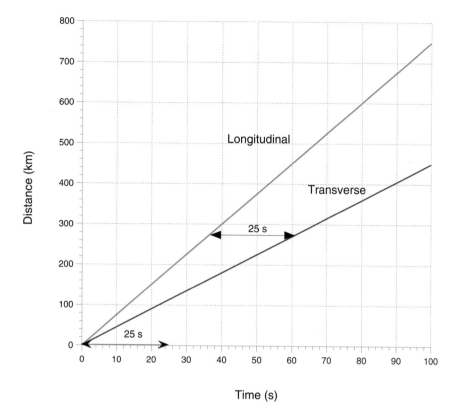

Figure 5.4
A segment representing a 25-s interval is drawn on a piece of paper and moved parallel to the horizontal axis until the horizontal distance between the two lines represents 25 s.

and blue lines represents 25 s (Figure 5.4). In our example, this occurs at a distance of 270 km from the source of the earthquake.

12. Suppose a transverse and a longitudinal wave from an earthquake are recorded 40 s apart. How far was the origin of the earthquake from this observation station?

13. Use a copy of Figure 5.3.

   a. Graph a third line representing the difference in the arrival times of the longitudinal and transverse waves at various distances from the origin of the earthquake.

   b. Find the difference in the arrival times of the two waves at a distance of 400 km from the origin of the earthquake.

   c. Find the distance to the origin of the earthquake when the difference in arrival times is 13 s.

14. Choose a point on the line for the transverse wave in Figure 5.3.

   a. Use the time and distance for this point to calculate the speed of the transverse wave.

INTERNET ACTIVITY
5.7 Locating an Earthquake

In the preceding section, you learned to find the distance of the origin of an earthquake from the location at which the quake was detected. The distance alone does not tell you where the origin is located. It may be anywhere on the circumference of a circle around the observation point. To locate the origin of an earthquake requires at least three observation points.

In this activity, you will find the location of the origin of an earthquake from simulated data to be found at this internet website:

http://vcourseware5.calstatela.edu/VirtualEarthquake/VQuakeExecute.html

The activity follows the same reasoning that we applied in the preceding section. However, there are some differences in notation. The longitudinal wave is called the "primary" or "P-wave," because it arrives at the detector first. The slower transverse wave is called the "secondary" or "S-wave."

The graphs of distance and time in the activity are inverted. The distance is plotted on the horizontal axis and the time is plotted on the vertical axis.

The activity contains the term "epicenter," which is the point on the surface of Earth directly above the origin or "focus" of the earthquake. Except for observation stations located very close to the epicenter, the distance to the origin and the distance to the epicenter differ by only a few percent.

16. Suppose that, in the *Virtual Earthquake* activity, you had only two stations reporting instead of three. What would you know about the location of the origin of an earthquake?

17. Figure A shows the change in air pressure at a sound detector caused by the sound of two pieces of wood clapped together.

   a. Assuming that the height of the wave is related to the strength of the sound, approximately how long did the sound of the clap last?

   b. What is the ratio of your answer in part (a) to the roundtrip time of the sound in Experiment 5.3, The Speed of Sound?

   c. Suppose that the answer in part (b) was 0.5. Could you have used the method of Experiment 5.3 to measure the speed of sound? Why or why not?

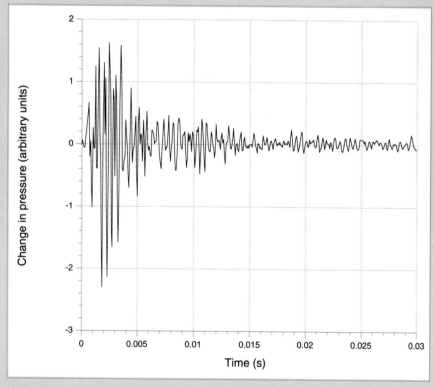

Figure A
For problem 17

18. Use your answer to part (a) of the preceding question and the speed of sound that your class measured to find the length, in meters, of the sound wave generated by the clap. In other words, how far had the leading edge of the sound wave traveled by the time the sound died out?

19. Suppose that a train pulls into a station at 4:45 P.M. and you know it has been traveling most of the time at a constant speed of 100 km/h.

a. Can you tell how far the train has traveled since it started earlier that day? Explain.

b. Suppose that an airplane flying along the tracks at 400 km/h overtook the train at some point. The plane flew over the station at 4:00 P.M. How far was the train from the station when the airplane overtook it? (*Hint:* Draw a graph of distance versus time for both the airplane and the train.)

THEME FOR A SHORT ESSAY

Write a science fiction story about a world in which weak sounds travel more slowly than strong ones. Include a description of some common experiences that would be different in such a world.

Heating and Cooling

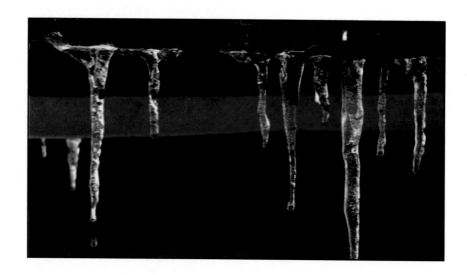

Chapter 6
Heating and Cooling

6.1 Introduction

Leave a cup of hot chocolate in a room and the hot chocolate will cool down. Leave a cup of cold juice on a table and the juice will warm up. In both cases, it appears that the temperature of the liquid changed without producing any other temperature changes or other noticeable effects on the surroundings. This result is not surprising because a cup is so much smaller than a room. Now suppose that when the juice warms up, the room does cool off ever so slightly. The change in the temperature of the room would be too small for you to notice or even measure with a very sensitive thermometer.

If we place liquids in a well-insulated container, such as a thermos, the cooling and warming are slowed down considerably. Observing hot and cold liquids and solids in a well-insulated container will show us how they affect each other. Most of the influence of the room will be eliminated. In this chapter, you will do experiments in a simple insulated container called a *calorimeter*, which is described in the next section.

Bringing an object in contact with another that is hotter or colder is not the only way to change its temperature. When the driver of a car slams on the brakes and comes to a screeching halt, you may smell burned rubber. Apparently, the tires warmed up a great deal without coming into contact with a still warmer object. However, this warming was accompanied by another change—the fast-moving car stopped.

Perhaps you have slid down a rope and burned your skin although the rope was not any warmer than your hands. In this case, there was also another change taking place. You ended up being below your previous position.

A hot flame from a wood fire is produced without being in contact with a still hotter object. However, the burning wood turns to ash, and some oxygen is consumed.

Table 6.1 lists some common processes in which changes in temperature are accompanied by a variety of other changes.

The second column in Table 6.1 lists either an increase or a decrease in temperature. The changes listed in the third column are quite different from each other, but each is accompanied by the change in temperature noted in the second column. We shall call any of the changes listed in either column a *change in energy*. A change in temperature indicates a change in *thermal energy*. In this chapter and the next one, you will investigate some energy changes like those mentioned in the third column of Table 6.1. You will begin with the simplest case: mixing warm and cool water.

Table 6.1 Some Common Processes with a Temperature Change

Process	Temperature change	Other changes
A hot piece of metal is placed in cold water.	The metal cools down.	The water warms up.
A hot piece of metal is placed in a slush of ice and water.	The metal cools down.	The temperature of the slush remains at 0°C, but some ice melts.
A car travels down a long, steep road at constant speed.	The brakes heat up.	The car loses elevation.
The brakes are applied to a fast-moving car on a flat road.	The brakes heat up.	The car slows down.

EXPERIMENT
6.2 Mixing Warm and Cool Water

What will be the temperature change when equal masses of water at different temperatures are mixed? Does the result depend on how large the equal masses are?

Begin with two samples of water of equal mass. One sample should be at room temperature and the other, between 30 °C and 35 °C. Measure the temperatures of both samples just before you mix them. By measuring the final temperature of the mixture, you can find the increase or decrease in the temperature of each sample.

The thermometers used in this experiment have been designed for high resolution between 10°C and 40°C. Examine the scale on the thermometer to convince yourself that the markings are 0.2°C apart.

There are three possibilities when the top of the liquid column is between two markings on the thermometer. To read the temperature when the top of the column is halfway between two markings, you should add 0.1°C to the lower marking. The result will be more precise than if you simply rounded to the nearest 0.2°C. When the top of the column is closer to the lower marking, adding 0.05°C to the lower marking will increase the precision of the measurement. Similarly, when the top of the column is closer to the higher marking, add 0.15°C to the lower marking. A magnifying glass can help you in this task.

Your teacher will assign a mass of water between 30 g and 50 g for you to use in this experiment. To save time in measuring out the assigned mass of water, you can use the fact that the density of water is 1.00 g/cm³. This allows you to measure the water in a graduated cylinder rather than massing it on a balance. (See Section 2.2.)

It is best to pour the cool water into a large test tube that is initially at room temperature. Place an equal mass of warm water at a temperature between 30°C and 35°C in the calorimeter (Figure 6.1).

Figure 6.1
A calorimeter construct-ed from two Styrofoam cups and a beaker. The beaker provides addi-tional insulation and prevents the calorimeter from tipping over. The test tube holds water at room temperature.

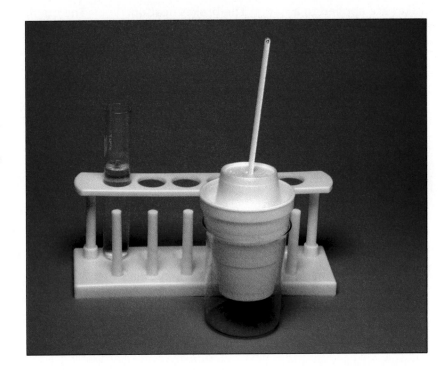

- Why is the procedure suggested above better than putting the cool water in the calorimeter and the warm water in the test tube?

- Which water sample's temperature should you measure first, the one at room temperature or the warm one? Why?

As soon as you record the initial temperatures of both samples, pour the cooler sample into the calorimeter and quickly replace the lid. Swirl the calorimeter in a circular motion to mix the samples. When the tempera-ture of the mixture remains constant for several seconds, record it.

- What is the temperature decrease for the warm-water sample?

- What is the temperature increase for the cool-water sample?

- What is the ratio of the decrease in temperature for the warm sample to the increase for the cool sample?

- How does your ratio compare with those of your classmates?

Part B

Does the ratio of the temperature decrease to the temperature increase depend on the mass ratio of the two samples of water? To find out if it does, repeat the experiment with a 60.0-g (60.0 cm^3) sample of warm water and a 30.0-g (30.0 cm^3) sample of cool water. As before, it is best to have the warm water in the calorimeter.

- What is the temperature decrease for the 60.0-g warm-water sample?
- What is the temperature increase for the 30.0-g cool-water sample?
- What is the ratio of the decrease in temperature for the warm sample to the increase in temperature for the cool sample?
- What is the ratio of the larger mass of water to the smaller mass?
- What is the relationship between these two ratios?

1. Why is it important to put the top back on the calorimeter immediately after adding the cool water to the warm water?

2. When mixing samples of different mass, why is it better to put the larger sample of warm water in the calorimeter and add the smaller sample of cool water to it?

3. Was it important to note the temperature readings on the thermometer before you measured the temperatures of the warm and cool water?

4. Why is an 8.5-oz (250 cm^3) Styrofoam cup filled with water more effective in maintaining a constant temperature than one containing only 50 cm^3 of water?

5. Suppose that instead of mixing 30 g of cool water and 30 g of warm water in the first part of this experiment, you mixed 90 g of each. Assuming that you started with the same initial temperatures, how would this change affect the final temperature of the mixture? How would it affect the change in temperature for each of the samples?

6.3 A Unit of Energy: The Joule

In the preceding experiment, you compared the ratio of masses with the ratio of temperature changes. You may have found the relationship between these two ratios on your own. If not, here are additional data taken with the same equipment that you used (Table 6.2).

We observe in every trial that the larger mass had a smaller temperature change. Furthermore, the ratios of the temperature changes are very close to the reciprocals of the ratios of the masses (Table 6.3).

The fact that these ratios are reciprocals of each other suggests that we examine the product of mass and temperature change for each water sample (Table 6.4).

Table 6.2 Temperature Changes for Unequal Masses of Water

Trial #	Mass of warm sample (g)	Decrease in temperature (°C)	Mass of cool sample (g)	Increase in temperature (°C)
1	90.0	3.25	30.0	9.70
2	80.0	2.05	20.0	8.05
3	100.0	1.75	20.0	8.75

Table 6.3 Ratios of Masses and Ratios of Temperature Changes for Data in Table 6.2

Trial #	Ratio of masses (warm/cool)	Ratio of temperature changes (decrease/increase)	Reciprocal of ratio of masses	Decimal equivalent of reciprocal of ratio of masses
1	3.00	0.335	1/3	0.333
2	4.00	0.255	1/4	0.250
3	5.00	0.200	1/5	0.200

Table 6.4 Mass times Temperature Change for Unequal Masses of Water

Trial #	Mass of warm water (g)	Decrease in temperature (°C)	Mass · (temperature decrease) (g ·°C)	Mass of cool water (g)	Increase in temperature (°C)	Mass · (temperature increase) (g ·°C)
1	90.0	3.25	293	30.0	9.70	291
2	80.0	2.05	164	20.0	8.05	161
3	100.0	1.75	175	20.0	8.75	175

The products of mass and temperature decrease (or increase) are shown in boldface type. In each of the three trials, these quantities are equal within the uncertainty of the measurements. We can now define the change in thermal energy to be proportional to the product:

$$\text{mass} \cdot (\text{change in temperature})$$

With this definition, the increase in thermal energy of the cool water equals the decrease in thermal energy of the warm water. We can make this statement even though we have not chosen a unit for energy. However, to say by how much the thermal energy of the cool water has increased, we must choose a unit for energy. The unit of energy we choose determines the value of the proportionality constant. (See Appendix 1 for help with proportionality.)

Quite a few different units of energy have been used over the years. The standard unit of energy in science today is the *joule* (J). To express the increase in thermal energy in joules, the proportionality constant must be expressed in J/g·°C. This constant is the amount of thermal energy (ThE), in joules, required to raise the temperature of one gram of water by one degree Celsius. It is called the *specific heat* of water.

$$\text{Change in ThE} = (\text{specific heat}) \cdot \text{mass} \cdot (\text{change in temperature})$$

The specific heat of water is 4.18 J/g·°C. The reason for choosing the joule as a unit of energy will be seen in the next chapter.

As an example, we calculate the decrease in thermal energy of the warm water in Trial 1. From Table 6.4, we know that the temperature of 90 g of water decreased by 3.25°C. The decrease in thermal energy in joules is:

$$\text{Decrease in ThE} = 4.18\frac{\text{J}}{\text{g} \cdot {}^\circ\text{C}} \times 90 \text{ g} \times 3.25 \, {}^\circ\text{C} = 1,223 \text{ J}$$

---

6. Use the data in Table 6.2 to verify the data in Table 6.3.

7. Suppose a freshly poured mug of tea contains 200 g of water at 75°C. If the tea cools down to room temperature (22°C), what is the decrease in thermal energy of the water?

8. The temperature inside a refrigerator at normal setting is 7°C. Suppose a 1.0-L bottle of water that has been in a refrigerator for some time is allowed to warm up to room temperature. What is the increase in the thermal energy of the water?

EXPERIMENT
6.4 Cooling a Warm Solid in Cool Water

When a warm piece of metal is put into water at room temperature, we expect the final temperature to be somewhere between the two initial temperatures. Suppose the masses of the metal and the water samples are the same. Will the final temperature of the two be in the middle, as it was when you mixed equal masses of water? To find out, you can put a metal washer in a sample of water of equal mass and measure the change in temperature for each.

Mass a large metal washer and tie a string to it. Submerge the washer in warm water between 30°C and 35°C, with the string hanging over the lip of the container for easy pickup (Figure 6.2).

• How do you know when the washer and the water are at the same temperature?

Measure a volume of room-temperature water that has a mass equal to the mass of the washer, and place it in your Styrofoam-cup calorimeter.

• Which temperature should you measure first, that of the metal washer or that of the water in the calorimeter? Why?

Measure the temperatures of both, then quickly remove the washer from the warm water and submerge it in the water in the calorimeter and stir gently.

• What are the increase in the temperature of the water and the decrease in the temperature of the washer?

• What is the increase in the thermal energy of the water?

Figure 6.2

The washer in the warm water in the beaker. Note the string that is used to lift the washer and put it in the calorimeter. The room-temperature water is in the calorimeter already. (The beaker to hold the calorimeter is not shown.)

The water and the washer in the calorimeter are quite insulated from the surroundings. Therefore, we set:

Decrease in ThE of metal = Increase in ThE of water.

The decrease in the thermal energy of the washer is given by:

Decrease in ThE of metal =
(specific heat of metal) · mass · (decrease in temperature).

- Using this relationship, find the specific heat of the metal.

9. Consider a 150.0-g piece of the same metal used to make the washer in Experiment 6.4. If the temperature of this piece is raised by 10.0°C, by how much will its thermal energy increase?

10. What reasons can you give for placing the sample of water, rather than the washer, in the calorimeter in Experiment 6.4?

6.5 Specific Heats of Different Substances

Your results in the preceding experiment may have surprised you, but they are quite valid. We did the same experiment, using the same equipment, with water and a piece of aluminum (Table 6.5).

Table 6.5 Data for Experiment 6.4 Using Warm Aluminum and Cool Water

Sample	Mass (g)	Initial temperature (°C)	Final temperature (°C)	Change in temperature (°C)
Aluminum	17.83	37.60	22.20	−15.40
Water	33.2	20.40	22.20	1.80

The gain of thermal energy by the water is:

$$(4.18 \text{ J/g} \cdot {}^{\circ}\text{C}) \times 33.2 \text{ g} \times 1.80{}^{\circ}\text{C} = 250 \text{ J}$$

This gain equals the loss of thermal energy of the piece of aluminum:

$$(\text{specific heat of aluminum}) \cdot 17.83 \text{ g} \times 15.40{}^{\circ}\text{C} = 250 \text{ J}$$

Or

$$\text{Specific heat of aluminum} = \frac{250 \text{ J}}{17.83 \text{ g} \times 15.40 \, {}^{\circ}\text{C}} = 0.91 \, \frac{\text{J}}{\text{g} \cdot {}^{\circ}\text{C}}$$

The specific heats for several substances are listed in Table 6.6. Notice that the specific heat of water is much higher than that of all the other common substances except hydrogen. This is generally true; very few substances have a specific heat as great as that of water.

Table 6.6 Specific Heats of Some Common Substances (to two significant digits)

Substance	Specific heat $(J/g \cdot {}^\circ C)$
Hydrogen	14
Water	4.2
Ice	2.1
Water vapor	2.0
Ethanol	2.5
Ethylene glycol	2.4
Olive oil	2.0
Air	1.0
Aluminum	0.92
Granite rock	0.79
Iron	0.44
Copper	0.38
Lead	0.13
Gold	0.13

11. Samples of copper and lead were each heated 5.0°C. What was the increase in the thermal energy per gram for each sample?

12. To raise the temperature of a sample of a metal from 20.0°C to 25.0°C required 12 J.

    a. How much thermal energy is required to raise the temperature of the sample by 1.0°C?

    b. What additional information must you have to determine the specific heat of the substance that makes up the sample?

EXPERIMENT
6.6 Melting Ice

When warm water is mixed with cool water, the final temperature of the water will be somewhere in between. What will be the final temperature when some warm water is added to a mixture of ice and water without melting all the ice?

To find out, pour about 25 g of water into a calorimeter and add two or three ice cubes. The ice cubes need not be completely covered with water. Stir the mixture until the temperature no longer changes.

- What is the temperature of the mixture?

Since you will be adding a known amount of warm water to this mixture, some of the ice will melt. Therefore you will want to know how much ice is in the calorimeter just before you add the warm water. Plan a strategy to mass the ice rapidly and return it to the calorimeter before much melting can take place. Having a balance nearby will help.

- What is the mass of the ice?

Now you are ready to add about 35 g of warm water (at about 40°C) to the mixture of ice and water. The exact amount is not important as long as you measure its mass and temperature accurately.

- What is the mass of the warm water?
- What is the temperature of the warm water?

Stir the mixture until the temperature no longer changes.

- What is the final temperature of the mixture?
- How much ice has melted?
- How much thermal energy was lost by the warm water?
- How much thermal energy was lost by the warm water to melt one gram of ice?

---

13. Why was it necessary to mass only the ice, but not the water, at 0°C before adding the warm water in the experiment?

6.7 Heat of Fusion and Heat of Vaporization

The results of the preceding experiment can be summarized as follows: in a calorimeter, some warm water cooled down and some ice melted to become water without a change in temperature. We mention the calorimeter to indicate that the water and ice were isolated from the room. Following the plan outlined in Section 6.1, we associate the melting with a form of energy, which is measured by the loss of thermal energy of the warm water. The gain in this energy when one gram of ice at 0°C becomes one gram of water at 0°C is called the *heat of fusion*. When one gram of water at 0°C freezes, it loses the same amount of energy. The surroundings gain an equal amount of thermal energy.

Table 6.7 lists the heat of fusion for a few substances. As you can see, the heat of fusion varies a great deal from substance to substance. The table also includes the *melting point* for each of these substances, that is, the temperature at which the transition from solid to liquid takes place.

Table 6.7 Heats of Fusion and Melting Points for Some Common Substances

Substance	Heat of fusion (J/g)	Melting point (°C)
Nitrogen	25.7	–210
Sulfur dioxide	115	–72.7
Ethanol	109	–117
Isopropanol	16.2	–89.5
Tin	60.9	231.7
Lead	24.7	327.3

Just as there is a form of energy associated with the melting of ice, there is also a form of energy associated with the boiling of water. To see this, think of an experiment similar to Experiment 6.4, Cooling a Warm Solid in Cool Water. Instead of cool water, we have water close to boiling. Instead of a washer, we have a stone at a temperature of several hundred degrees Celsius. When the stone is dropped into the water, the water will boil and the stone will cool down. However, a thermometer will show that the temperature of the boiling water remains constant (100°C at sea level). Thus, water evaporates while the temperature of the stone decreases. The form of energy associated with the evaporation of one gram of water at the boiling point is called the *heat of vaporization*. To measure this energy, you would need to find the decrease in thermal energy of the stone.

Table 6.8 lists the heat of vaporization for several common substances along with the *boiling point* of each, that is, the temperature at which the transition from liquid to gas takes place.

Table 6.8 Heats of Vaporization and Boiling Points for Some Common Substances

Substance	Heat of vaporization (J/g)	Boiling point (°C)
Nitrogen	200	–196
Sulfur dioxide	383	–10.0
Ethanol	855	78.5
Water	2,260	100
Isopropanol	669	82.4
Tin	2,490	2,260
Lead	859	1,740
Helium	20.9	–269

14. How do the values for the heat of fusion in Table 6.7 compare with the value you found for ice in Experiment 6.6, Melting Ice?

15. Suppose that it rains when the temperature of the ground, and the air just above the ground, is below the freezing point of water (0°C). The rain freezes. Will the freezing lower or raise the temperature near the ground? Explain.

16. How does the heat of vaporization of water compare with those of the other substances listed in Table 6.8?

17. How do the heats of fusion and vaporization of the same substances compare?

FOR REVIEW, APPLICATIONS, AND EXTENSIONS

18. Suppose you heat 100 g of room-temperature water in a small beaker and 1,000 g of room-temperature water in a large beaker until the temperature of each sample is 60°C. Do both water samples gain the same amount of thermal energy? Explain your reasoning.

19. Figure A shows the cooling curves for four 50-g samples of water. The temperature of each water sample was recorded as a function of time, and then graphed on the same set of axes.

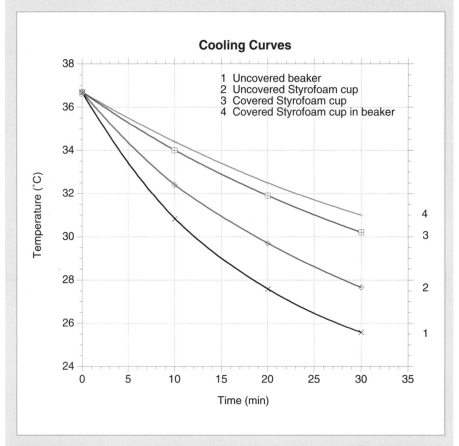

Figure A
For Problem 19

a. Which arrangement makes for the best calorimeter? Why?

b. What was the approximate decrease in temperature for each sample during the first 2 min?

c. What was the approximate decrease in temperature for each sample during the last 2 min?

d. Room temperature was about 22°C. As the temperature of the water approached room temperature, was the decrease in temperature in a 2-min interval greater than or less than at the beginning?

e. Why were the experiments in this chapter carried out close to room temperature?

20. Suppose the volume of water in a swimming pool is approximated by a rectangular box of length 9.6 m, width 4.8 m, and average depth 1.7 m.

   a. What is the volume of the water in cubic meters? in cubic centimeters?

   b. What is the mass of the water in grams?

   c. Suppose the pool's owner wants to raise the temperature of the water by 5°C. By how much will the thermal energy of the water increase in joules, neglecting any loss of thermal energy to the air and the walls of the pool?

   d. Swimming pools are usually heated electrically. The electric company uses the kilowatt-hour (kW·h) as the unit of energy in its billings. 1 kW·h = 3.6×10^6 J. What is the needed increase in thermal energy in kW·h?

   e. Check your family's electricity bill for the current price of a kilowatt-hour. How much would it cost to heat this swimming pool?

21. Equal masses of olive oil and water are heated in identical containers for the same length of time on the same hot plate. If the temperature of the water increases by 5°C, what will be the approximate temperature change of the olive oil?

22. Some metal washers, whose total mass is 200 g, are heated in boiling water and then placed in 100 g of water at 20°C. The final temperature of both is 25°C.

   a. What is the specific heat of the metal?

   b. The washers might be made of which metal?

23. The specific heat of zinc is 1.62 J/g·°C. How much thermal energy is required to raise the temperature of 4.00 g of zinc by 2.0°C?

24. One-hundred-gram samples of aluminum, copper, and lead are placed in boiling water. Each sample is then removed and placed in a container with 100 g of water at 20°C. Which metal will cause the greatest temperature change in the cool water? (See Table 6.6.)

25. You can put your hand into a hot oven for a moment without harming yourself. Yet holding your hand over steam at 100°C can cause a serious burn. Why?

26. Suppose that 1 g of ice at 0°C is changed into 1 g of steam at 100°C. What is the total increase in energy?

27. A century ago farmhouses in colder parts of the country often had unheated cellars for storing vegetables. The vegetables were kept from freezing by placing large buckets of water in the cellar. Explain how the water helped prevent the vegetables from freezing.

28. Eggs will cook just as quickly at a low boil as at a high boil. Why?

THEME FOR A SHORT ESSAY

Suppose a friend admits to being confused about the difference between temperature and thermal energy. Write a short essay using examples from daily life that will highlight the distinction between these two terms.

Chapter 7

Potential Energy and Kinetic Energy

EXPERIMENT
7.1 Heating Produced by a Slowly Falling Object

In the previous chapter, you studied changes in temperature accompanied by other changes. In Experiment 6.2, Mixing Warm and Cool Water, and in Experiment 6.4, Cooling a Warm Solid in Cool Water, an increase in temperature was accompanied by a decrease in temperature. In Experiment 6.6, Melting Ice, a decrease in temperature was accompanied by melting of ice.

In all those experiments, the objects involved did not change their position in any way. In the experiment in this section, a heavy object held by a nylon line will fall very slowly. The nylon line will rub against an aluminum cylinder, raising its temperature (Figure 7.1). The purpose of this experiment is to find out how the increase in thermal energy of the cylinder is related to the weight of the falling object. This knowledge will provide a way to define a new kind of energy. To obtain reliable results, you will have to pay attention to several details, as described below.

A thermometer will be inserted in a hole in the cylinder. It is important to establish good thermal contact between the thermometer bulb and the cylinder. This can be most easily achieved by filling the cylinder hole with a conducting paste. Before you put the cylinder in place, be sure you know its mass, including the paste. Then clamp the cylinder firmly between two blocks of wood so that it cannot rotate. Finally, insert the thermometer gently as far as it can go. Turn the thermometer to a position where it can be read most conveniently.

Once the thermometer is in place, wrap the nylon line around the cylinder and attach it to a water-filled container or some other heavy object. Hook a small counterweight to the free end of the nylon line. The counterweight will hold the line taut so that it rubs against the aluminum cylinder as the object falls.

Figure 7.1

Apparatus for measuring the heating produced by a slowly falling object. The nylon line supporting the water-filled container is looped around a fixed aluminum cylinder. The line then passes down over a steel guide-rod on the left and is anchored with a counterweight hanging from the free end.

The force of friction between the line and the cylinder depends on the number of times the line is wound around the cylinder. With too many turns, the friction will be greater than the weight of the hanging object, so the object will not fall. With too few turns, the object will fall too rapidly, so the friction will be too low and the cylinder will not warm up.

Before performing the experiment, you will need to determine how many times to wrap the nylon line around the aluminum cylinder so that the object takes between 5 s and 15 s to complete its fall. Begin by wrapping four or five turns of nylon line around the cylinder. As the hanging object slowly falls, the loops of nylon line will creep along the cylinder. To provide enough distance, wind the nylon line around the cylinder in a clockwise direction. Hang a washer on the line to keep the line taut until the heavy object is attached (Figure 7.2(a)). Slide the loops along the cylinder so that they form a single layer and are pressed up against the rim at the front of the cylinder (Figure 7.2(b)).

Now you are ready to attach the heavy object and time its fall. If necessary, adjust the number of turns of nylon line until the object falls at a satisfactory rate. Once you have determined the proper number of turns, mark the line where it begins to wrap around the cylinder, as shown by the arrow in Figure 7.2(b). Now you will be able to repeat the experiment with the heavy object released from the same height.

To study the effect of the net falling weight on the increase in thermal energy, you must keep all other variables as constant as possible. Specifically, you need to make sure that the object always falls the same distance.

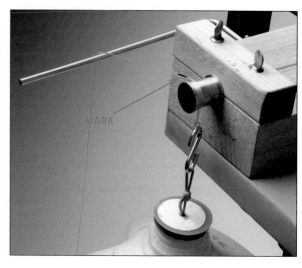

Figure 7.2

(a) The apparatus before the first run. The thermometer projects from the rear of the cylinder and the washer is in place.

(b) A close-up of the apparatus at the start of a run. The loops have been pulled to the front rim of the cylinder and the heavy weight has been attached.

Your teacher will tell you what height to use and will assign a weight to each group.

As the object falls, the counterweight is lifted. To determine the net falling weight, you must subtract the weight of the counterweight from the total weight of the object. If you are using a plastic container filled with water, the object's weight is the combined weight of the empty container and the water.

To find the weight of the empty bottle and the counterweight directly, you can use a spring scale. Alternatively, you can mass these objects and multiply the mass by the proportionality constant g. (See Section 2.1.) The weight of the water is most conveniently found by measuring its volume and using its density to calculate its mass. (See Section 2.2.)

- Why do you *not* need to be concerned about the weights of the identical hooks on the ends of the nylon line?

There is one other quantity that must be considered before you can perform the experiment — the amount of thermal energy lost to the surrounding air. In Chapter 6, you used a calorimeter to reduce this loss. But the aluminum cylinder is not in a calorimeter. As the nylon line heats the cylinder, a significant amount of the thermal energy could be lost to the room. To prevent this, you should pre-cool the aluminum cylinder so that it is slightly below room temperature at the beginning of each run and slightly above room temperature at the end of each run. In this way, gain and loss of thermal energy to the room will roughly balance. Cooling the cylinder by about 0.5°C to 0.7°C below room temperature will be sufficient for the object you will be using.

Before attaching the object to the apparatus, practice pre-cooling the object several times with an ice cube in a plastic bag. If needed, warm the cylinder with your hand to bring it up to the desired temperature.

With the nylon line in place and the cylinder cooled correctly, attach the object and make a measured run. Repeat the experiment several times. From the temperature changes, the mass of the cylinder, and the specific heat of aluminum, calculate the increase in thermal energy of the cylinder for each run.

Using the data from the entire class, graph the increase in thermal energy as a function of the net falling weight.

- Why should you consider the origin as a data point?
- How is the amount of thermal energy gained by the cylinder related to the net falling weight?

1. **Suppose you arrange the apparatus in Experiment 7.1 so that it takes the object 10 min to reach the floor. What rise in temperature do you expect to find?**

2. Suppose that the aluminum cylinder had twice the mass of the cylinder that you used. How would this affect the rise in temperature in your experiment?

3. How would your experimental results have been affected by each of the following?

   a. You began with the temperature of the aluminum cylinder a degree or two above room temperature.

   b. You forgot to subtract the weight of the counterweight from the weight of the container.

4. How would your experimental results have been affected if you had used the same volume of candle oil instead of water?

7.2 Gravitational Potential Energy

The preceding experiment showed that the increase in thermal energy of the aluminum cylinder is related to the weight of the object changing its vertical position by a given distance. The combined results for the class showed that the increase in thermal energy of the cylinder is proportional to the net weight of the falling object, keeping the falling distance constant.

How does the increase in thermal energy depend on the falling distance? To answer this question, we have to vary the falling distance while keeping the weight constant. Figure 7.3 presents data obtained from just such an experiment carried out from a greater height than the one you used. The weight of the object was 50.0 N.

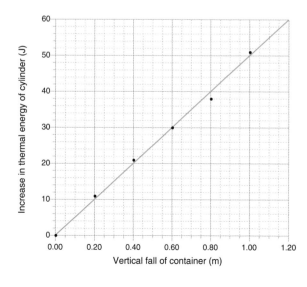

Figure 7.3
A graph of the increase in thermal energy of the cylinder as a function of the distance that a 50.0-N object has fallen.

We now have two proportions. (1) If the falling distance is kept constant, the increase in thermal energy is proportional to the weight of the object. (2) If the weight is kept constant, the increase in thermal energy is proportional to the distance. To satisfy both these results, the increase in the thermal energy of the cylinder must be proportional to the product of the weight and the falling distance. (See Appendix 1 Proportionality.)

How can we find the proportionality constant for this relationship? We take any point on the graph in Figure 7.3 and divide the increase in the thermal energy of the cylinder by the distance fallen. For example, a 1.00-m fall of the 50-N object increased the thermal energy (ThE) by 50.0 J. For a 1-N object, therefore, the increase would be just 1.00 J. This means that the proportionality constant is 1.00 J/N·m.

This result is not a coincidence. Historically, one joule was defined as one newton-meter.

$$1\,J = 1\,N \cdot 1\,m$$

This definition makes the proportionality constant a pure number (a number without units). A pure number of value one need not be included in the equation.

$$\text{Increase in ThE (J)} = \text{weight (N)} \cdot \text{(distance fallen) (m)}$$

Note that the left side of the above equation depends only on the cylinder: its mass, its substance, and its temperature change. The right side of the equation depends only on the falling object: its weight and the distance it falls. We now use this equation to define a new form of energy. Because it depends on the force of gravity, it is called *gravitational potential energy* (GPE):

$$\text{Decrease in GPE (J)} = \text{Increase in ThE (J)}$$

Figure 7.4

The same graph as in Figure 7.3 but with changed labels on the axes. The graph shows the increase in gravitational potential energy of a 50.0-N object as a function of the distance it is raised.

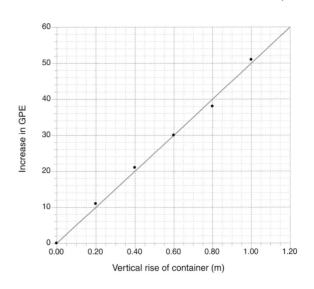

By equating the increase in thermal energy of the cylinder with the decrease in gravitational potential energy of a falling object, the sum of the two energies stays constant. It also means that when an object rises, its GPE increases.

$$\text{Increase in GPE (J)} = \text{weight (N)} \cdot \text{(distance raised) (m)}$$

In fact, we can use Figure 7.3 to describe the change in GPE of the 50-N object by simply changing the labels on both axes (Figure 7.4).

5. A one-gallon bottle of water weighs about 40 N. How much gravitational potential energy does the bottle gain if it is lifted each distance?

   a. 1 m   b. 2 m   c. 10 m

6. Which gains more gravitational potential energy, a 4-N object lifted 3.0 m or a 5-N object lifted 2.0 m?

7. Suppose that you repeated Experiment 7.1 in a laboratory on the moon, using the same equipment that you used on Earth. If the falling object dropped the same distance as it did on Earth, what would be the temperature rise of the aluminum cylinder?

7.3 Elastic Potential Energy

We can show the existence of another common form of energy by the following demonstration with a spring that is fixed at one end. When the free end of the spring is pulled and then suddenly released, the spring will snap back. Suppose that the free end of the stretched spring is attached to a nylon line that is wrapped around an aluminum cylinder and has a counterweight at its other end (Figure 7.5). The pull of the spring can be almost balanced by the friction with the cylinder. The spring will then con-

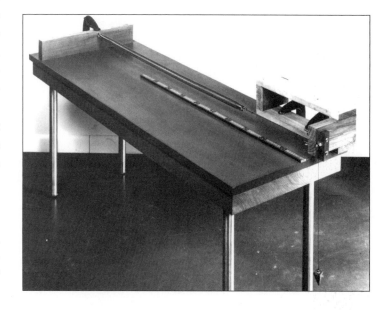

Figure 7.5
The free end of a stretched spring pulls a nylon line wrapped around an aluminum cylinder. When the spring contracts, the temperature of the cylinder rises.

tract slowly, pulling the nylon line along with it and raising the temperature of the cylinder. The increase in thermal energy of the cylinder is determined by the rise in temperature of the cylinder.

This demonstration shows an increase in thermal energy accompanied by the contraction of a stretched spring. When we stretch the spring to different initial lengths and allow it to contract to the same final length, we find different rises in temperature (Table 7.1).

Table 7.1 Data from Stretched Spring Experiment

Position of free end of spring (m)		Amount of contraction (m)	Increase in temperature (°C)	Increase in thermal energy (J)
Initial	Final			
0.50	0.20	0.30	0.75	10
0.60	0.20	0.40	1.10	14
0.70	0.20	0.50	1.55	20
0.80	0.20	0.60	2.10	27
0.90	0.20	0.70	2.70	34
1.00	0.20	0.80	3.40	43

Now let's compare the change in position of the free end of the spring with the change in position of the falling object discussed in the preceding section. In both cases, a change in position of one object is accompanied by an increase in the thermal energy of another. In the case of the slowly falling object, we defined a decrease in gravitational potential energy in terms of the increase in thermal energy of the cylinder. We now define a decrease in *elastic potential energy* in the same way: The decrease in the elastic potential energy of the contracting spring is defined as equal to the increase in the thermal energy of the cylinder.

Decrease in elastic potential energy = Increase in ThE

As in the case of gravitational potential energy, the sum of the two energies does not change as the spring contracts. A stretched spring has a higher elastic potential energy than a spring with no tension.

Note that all the experiments discussed so far in this chapter and in Chapter 6 give information about only changes, that is, increases and decreases, in energy. None of the experiments can be used to answer the question, "What is the thermal energy of 1 L of water at 20°C?" or, "What is the gravitational potential energy of a book on the table?" We can assign a value of zero to the gravitational potential energy of the book when it is on the table or when it is on the floor. It is a matter of convenience. The same is true for elastic potential energy. It is convenient to assign a value of zero elastic potential energy to the spring at its natural length.

8. In Figure 7.5, the distance that the spring contracts determines the increase in thermal energy of the cylinder.

   a. Using the data in Table 7.1, graph the increase in thermal energy versus the contraction of the spring. Include the origin of the coordinates as a data point (no contraction, no increase in thermal energy).

   b. How should you re-label the axes of your graph to show how the elastic potential energy of the spring depends on the stretch?

9. How do the graphs you drew for Problem 8 (a) and (b) compare with Figures 7.3 and 7.4?

7.4 Kinetic Energy

Have you ever wondered what would happen if you tried to stop a rapidly spinning bicycle wheel by holding your hand against the tire? You should not try this because your hand would heat up as the wheel slowed down. In fact, your hand might even be burned. An increase in thermal energy occurs although the wheel has neither fallen nor been lifted. This happens whether the wheel spins in a vertical or a horizontal plane, so there is no change in the gravitational potential energy of the wheel. Therefore, the heating up of your hand must be related to the slowing down of the wheel. Once again, as in Section 6.1, we define another new form of energy. This form of energy, which is associated with motion, is called *kinetic energy*.

What does kinetic energy depend on? As we have already pointed out, gravity and, therefore, the weight of the wheel have nothing to do with it. Imagine the same bicycle wheel is spinning at the same rate on the moon, where the wheel would weigh only about one sixth of its weight on Earth. If you stopped the wheel with your hand, your hand would heat up exactly the same amount as it would on Earth. That is, the thermal energy of your hand would increase by the same amount.

Now suppose you stopped two identical wheels spinning on the same axis. Together, the wheels have twice the mass of one wheel, and so they will double the increase of thermal energy of your hands. Apparently, kinetic energy is proportional to the mass of a moving object. Mass, as measured by an equal-arm balance, does not change with location (Section 2.1).

Figure 7.6

A spinning wheel. The black rim is a steel ring covered with black tape. A metal pointer fastened to the stand is lightly tapped by a piece of tape attached to one of the spokes each time that spoke passes. The sound produced by the tape hitting the pointer is used to count the time for 10 rotations.

POINTER

Figure 7.7

A close-up of the center of the wheel and the aluminum cylinder just before it is inserted into the plastic extension of the hub.

How does kinetic energy depend on speed? This question can be answered by experiment. Figure 7.6 shows a rapidly spinning bicycle wheel. The tire has been replaced by a solid ring of steel, about 1.5 cm thick. This effectively shifts the entire mass of the wheel (4.60 kg) to the rim. A plastic extension of the hub contains a depression into which an aluminum cylinder can be inserted (Figure 7.7). After the wheel has been

spun and its rate of spinning determined, the wheel is brought to rest by friction between the hub and the cylinder. The rise in the temperature of the cylinder is read on a thermometer that is like the one you used in Experiments 7.1 and 7.6 (Figure 7.8).

The data and calculated values for this experiment are displayed in Table 7.2.

Table 7.2 Data from Spinning Wheel Experiment

Time for 10 rotations (s)	Initial speed of rim (m/s)	Increase in temperature (°C)	Increase in thermal energy (J)
15.4	1.23	0.30	4.0
13.0	1.45	0.30	4.0
10.8	1.75	0.45	6.1
10.6	1.78	0.50	6.8
10.0	1.89	0.65	8.8
10.0	1.89	0.50	6.8
9.7	1.95	0.65	8.8
8.6	2.20	1.00	13.5
8.0	2.36	1.00	13.5
7.0	2.70	1.05	14.2
6.6	2.86	1.55	20.9
6.4	2.95	1.30	17.6

The data in the first and third columns show the quantities that were measured directly. The entries in the second and fourth columns are calculated.

The speed of the rim was calculated by dividing the circumference of the rim by the time of one rotation:

$$\text{Speed} = \frac{\pi \cdot \text{diameter}}{\text{time of one rotation}}$$

If the time of one rotation is doubled, the speed is cut in half. The speed is inversely proportional to the time of one revolution. This is why the entries in the first column are in descending order, while those in the second column are in ascending order. The diameter, measured from the middle of the steel rim, is 0.603 m.

The increase in thermal energy in the fourth column was calculated by multiplying the increase in temperature by the product of the mass of the cylinder and its specific heat. For the cylinder used in this experiment, the value of this product is 13.5 J/°C.

Figure 7.9 contains a graph of the data in the second and fourth columns of Table 7.2. The origin was added to the data because we can be absolutely certain that a wheel at rest will cause no increase in thermal energy. Two features of the graph are evident: the data points are quite scattered, and the smooth curve is not a straight line.

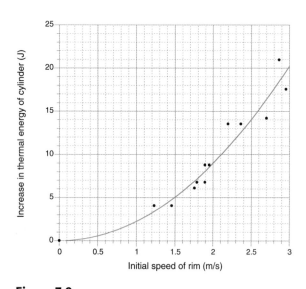

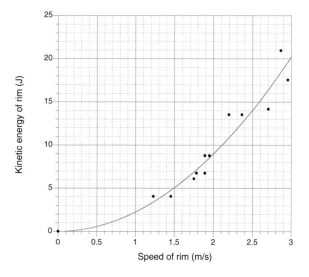

Figure 7.9

(a) A graph of the increase in thermal energy of the cylinder as a function of the initial speed of the rim. The point at the origin was added to the values in Table 7.2.

(b) The same graph with the axes renamed.

10. Confirm the first and last entries in the second and fourth columns in Table 7.2. Use the information given in the text about the diameter of the wheel and the product of the mass and specific heat of the aluminum cylinder.

11. Look at the fourth row of Table 7.2. The temperature increase of 0.50°C came from an initial temperature of 21.35°C and a final temperature of 21.85°C.

   a. Another experimenter might have read the initial temperature as 21.30°C. Assuming that the final temperature reading was unchanged, what would be the increase in the thermal energy of the cylinder?

   b. How would this reading affect the position of this data point in Figure 7.9?

   c. The time of 10 rotations was measured with a stopwatch. The reading of 9.7 s might have been off by 0.1 s either way. How would a reading of 9.6 s or 9.8 s affect the position of the data point in Figure 7.9?

   d. The reading of which quantity, the time or the temperature, has a greater effect on the position of the data points?

7.5 Kinetic Energy as a Function of Speed

As we have already seen, the graph in Figure 7.9(b) is not a straight line. Is there a simple way of expressing the relationship between speed and kinetic energy?

The graph is a good representation of all the data points, considering their uncertainties, so let's look more carefully at Figure 7.9(b).

At 1.00 m/s the kinetic energy of the wheel is 2.2 J. At 2.00 m/s the wheel has a kinetic energy of 8.8 J, or four times the kinetic energy at 1.00 m/s. At 3.00 m/s the kinetic energy is 20 J, or nine times the value at 1.00 m/s. This suggests that we calculate the squares of the speeds and plot kinetic energy as a function of the square of speed. The resulting graph is shown in Figure 7.10.

Indeed, the straight line through the origin shows that the kinetic energy of the rim is proportional to the square of the speed. We already know that kinetic energy is proportional to the mass of a

Figure 7.10
A graph that shows the relationship between the square of the speed of a wheel and its kinetic energy.

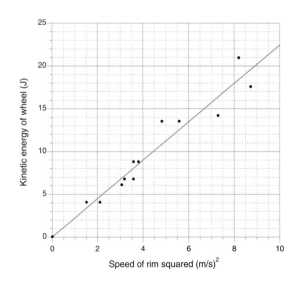

moving object. Therefore, kinetic energy (KE) must be proportional to the product of the mass and the square of the speed. That is,

$$KE = (\text{proportionality constant}) \cdot (\text{mass}) \cdot (\text{speed})^2.$$

We can now find the proportionality constant from the graph shown in Figure 7.10. At 10 (m/s)² the kinetic energy of the wheel is 22.5 J. The kinetic energy per (m/s)² of the same wheel will be

$$22.5 \text{ J}/10 \text{ (m/s)}^2 = 2.25 \text{ J/(m/s)}^2.$$

The mass of the wheel, which is almost all in the rim, is 4.60 kg. Therefore, the kinetic energy per kilogram per (m/s)² will be

$$\frac{2.25 \text{ J}/(\text{m}/\text{s})^2}{4.60 \text{ kg}} = 0.49 \frac{\text{J}}{\text{kg} \cdot (\text{m}/\text{s})^2}.$$

More accurate experiments show that the constant of proportionality is 0.500 for a wide range of speeds and masses. It is, therefore, generally written as the fraction $\frac{1}{2}$ and not as 0.500. To sum up, the kinetic energy of any object is given by the following formula:

$$KE = \tfrac{1}{2} \cdot (\text{mass}) \cdot (\text{speed})^2$$

If the mass is given in kilograms and the speed in meters per second, the kinetic energy is given in joules.

12. A compact car has a mass of about 1,000 kg.

   a. What is its kinetic energy at each of the following speeds? 0.5 m/s, 10 m/s, and 30 m/s

   b. When would a car be moving at these speeds?

13. Can two people be moving at different speeds and have the same kinetic energy?

14. Can two moving cars of different mass have the same kinetic energy?

15. Consider a cart moving at constant speed. What will increase its kinetic energy more, tripling its mass or tripling its speed?

EXPERIMENT
7.6 Free Fall

When a ball falls freely, it loses height but gains speed. Or, in other words, the ball loses gravitational potential energy and gains kinetic energy. Specifically, the loss in gravitational potential energy is

$$\text{Decrease in GPE} = \text{weight} \cdot (\text{distance fallen})$$

For a moving object, the kinetic energy is:

$$KE = \tfrac{1}{2} \cdot (\text{mass}) \cdot (\text{speed})^2$$

We arrived at both of these relationships by experiments with thermal energy. Except for some negligible heating due to friction with the air, there are no changes in thermal energy in free fall. Is the loss of gravitational potential energy equal to the gain in kinetic energy as the ball falls?

To answer this question, you will do an experiment similar to Experiment 4.5, Terminal Speed, but instead of dropping a coffee filter, you will drop a softball on a motion detector inside a protective cover (Figure 7.11).

The motion detector is sensitive to any motion and any object within the range of its beam. Therefore, it should not be set up near chairs, counters, or other objects that it might detect. Release the ball with minimum motion of the hand and arm, and do not move until the ball has stopped falling.

Figure 7.11
The softball just before its release. Notice the position of the student's arms and the position of the motion detector.

The software and the computer have a slight lag time after the "collect" button has been clicked on the computer. Therefore, the person operating the computer should click on "collect," wait until the graph trace reaches 1.0 s, and then say, "Go."

Make a few runs to become familiar with the procedure. The graph of a successful run will have three distinct parts (Figure 7.12). It will begin with a horizontal line for about 1 s, corresponding to the height at which the ball is being held. The second part will be a smooth downward curve until about 0.5 m, the shortest distance that the motion detector can measure. The third part will be quite irregular, depending on what the detector sees at the end of the run. This last part is of no interest in this experiment.

Suppose you want to know how much gravitational potential energy the ball has lost at some point in its fall. For example, consider the point that is 0.300 m below its release point.

- What is the loss in GPE of the ball when it is 0.300 m below the point of release?
- Where is this point on the graph shown in Figure 7.12?

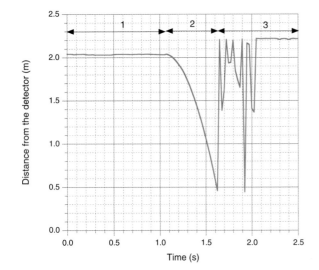

To find the gain in kinetic energy at this point, you must know the speed of the ball at that point. The only condition under which you learned to find a speed from a graph is when the speed is constant. You will recall that a speed is constant if the average speed is the same during any time-interval (Section 4.4). Since the graph is not a straight line, the average speed will vary depending on the interval for which it is calculated. However, if you look at a very short segment of the graph, it appears to be nearly a straight line. Figure 7.13 is an enlargement of part of Figure 7.12 in the vicinity of 1.70 m, which is 0.300 m below the point of release. Using a ruler, you can convince yourself that for any short segment, the graph is practically straight. You can, therefore, use the average speed over such a short time-interval as the speed at a given point. This speed is called the *instantaneous speed*.

Figure 7.12
A sample of the kind of distance-versus-time graph that may result from a freely falling softball.

- What is the speed of the ball at 0.300 m below the release point in your run? (An enlarged section of your graph similar to Figure 7.13 will be helpful.)
- What is the gain in kinetic energy of the ball at that point?
- What is the ratio of the gain in kinetic energy to the loss in gravitational potential energy at that point?

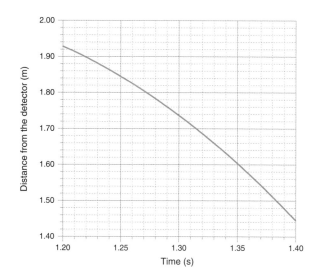

Figure 7.13

An enlargement of a section of Figure 7.12 between 1.20 s and 1.40 s. In intervals of 0.05 s, the line is almost straight.

Repeat the comparison of the loss in gravitational potential energy and the gain in kinetic energy at 0.600 m below the point of release.

• What is the ratio at this point?

---

16. **Use a copy of Figure 7.13.**

   a. **Calculate the average speed of the ball between 1.30 s and 1.35 s.**

   b. **Suppose that the ball moved at a constant speed equal to the average speed that you found in part (a). Draw a graph that represents the motion of a ball at that constant speed between 1.30 s and 1.35 s.**

   c. **Use the graph that you drew in part (b) to determine whether the average speed of the ball between 1.31 s and 1.34 s would differ from the average speed between 1.30 s and 1.35 s.**

7.7 The Law of Conservation of Energy

The results of the preceding experiment show that the definitions of GPE and KE are useful. They enable us to predict the speed of a freely falling object at any distance from the point of release. Conversely, we can predict how high a ball will rise when it is thrown upward with a known initial speed. The initial speed determines the initial kinetic energy of the ball. The ball will rise until it has lost all its kinetic energy and gained an equal amount of gravitational potential energy. The gain in gravitational potential energy determines how high the ball will rise.

The exchange of gravitational potential energy and kinetic energy does not have to take place with the same object. Figure 7.14(a) shows the same wheel pictured in Figure 7.6. Now a heavy object hangs from a string wrapped around the hub of the wheel. When the wheel is released, the string unwinds and the falling object causes the wheel to turn. Clearly, the falling object loses gravitational potential energy and gains some kinetic energy. However, the rim of the wheel, which is much heavier than the falling object, moves much faster. So the wheel must have gained a lot more kinetic energy!

In Figure 7.14(b), the object is momentarily at rest at its lowest point, and the wheel spins the fastest. At this instant, the spinning wheel begins to slow down as it pulls the object back up.

In Figure 7.14(c), the object and the wheel are again at rest. Notice that the object is at almost the same height from which it was released. Had we measured the height and speed of the object and the speed of the rim, we would have found that the loss in gravitational potential energy was nearly equal to the gain in kinetic energy at any time during the period of motion. Some loss is inevitable due to friction in the bearings and with the air.

A similar experiment is shown in Figure 7.15. The object has been replaced by a stretched spring. In Figure 7.15(a), the spring is stretched all the way to the rim of the wheel, as shown by the small ring at the end of the spring. In Figure 7.15(b), the spring has contracted to its natural length and the wheel spins the fastest. At this moment, the wheel begins to

Figure 7.14

(a) A string holding a heavy object is tied around the hub of the wheel. When the wheel is released, the object begins to fall and the wheel then begins to turn. (b) The string is completely unwound. The object is momentarily at rest at its lowest point. The wheel turns at its highest speed. (c) The wheel and the object are again at rest. The object is almost back to its starting position.

(a)   (b)   (c)

stretch the spring. In Figure 7.15(c), the wheel is again at rest and the spring is almost back to its initial length.

The experiments described in this section are just two examples of gains and losses of different forms of energy that always balance. Many other experiments lead to the same conclusions: when all changes are taken into account, the losses equal the gains. Saying it differently, the sum of all forms of energy involved in any process remains the same. This generalization is called *the law of conservation of energy.*

No process has ever been observed in which the law of conservation of energy has been violated. However, not every process that satisfies the law will necessarily take place. Think of Experiment 6.2, Mixing Warm and Cool Water. We defined the cooling of the warm water as a loss of thermal energy, so that the loss equaled the gain of thermal energy by the cool water. Imagine the opposite process. A quantity of water at some uniform temperature inside an insulated container separates by itself into two parts, one warmer and one cooler. This process never happens, although it would not violate the law of conservation of energy.

In Experiment 7.1, Heating Produced by a Slowly Falling Object, the aluminum cylinder gained thermal energy and the falling object lost gravitational potential energy. You will never see the cylinder cooling while the object is rising, except in a movie run backward. The law of conservation of energy tells us what is not possible. However, it is not sufficient to tell us what is possible.

(a)   **(b)**   **(c)**

Figure 7.15
A stretched spring contracts and turns a wheel. The spring is stretched by the turning wheel until it returns almost to its original position when the wheel comes to rest.

17. When it falls, the heavy object in Figure 7.14 never reaches the floor. Describe the energy changes of the object and the wheel until all motion stops.

18. Figure A shows an object hanging from a spring. In (a), the object is also supported by a block of wood. When the block is removed, the object falls. In (b), the object has reached its lowest position. Its next highest position is shown in (c).

   a. Describe the energy changes taking place between (a) and (b), and between (b) and (c).

   b. Consider an instant between (a) and (b). Will the decrease in gravitational potential energy of the object equal the increase in elastic potential energy of the spring?

   c. Based on the sequence of photographs, do you think the object will continue to bounce up and down forever?

   d. Does your answer to part (c) violate the law of conservation of energy?

19. A skydiver will fall with increasing speed at first but will soon reach a terminal speed of about 100 mi/h. How can you explain the fact that after reaching a constant speed, a skydiver will continue to lose gravitational potential energy without gaining any more kinetic energy?

Figure A
For Problem 18

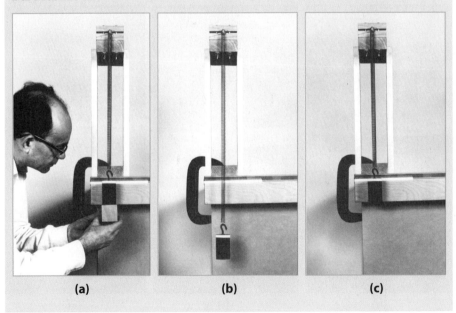

(a)   (b)   (c)

FOR REVIEW, APPLICATIONS, AND EXTENSIONS

20. In Figure B, a cart loaded with bricks can roll down a slanted board. Like the falling object in Experiment 7.1, Heating Produced by a Slowly Falling Object, the cart is tied to a nylon line wrapped around an aluminum cylinder. Also as in that experiment, a thermometer has been inserted into the cylinder.

Figure B
For Problem 20

The experimenter can change the angle at which the board is tilted. When the board is less steep, the cart travels a longer distance for the same decrease in height (Figure C).

Figure C
For Problem 20

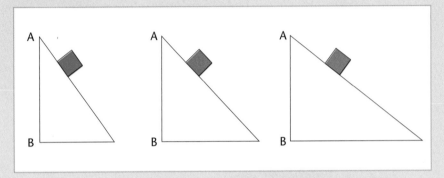

The table below shows the results of several runs.

Run number	Distance cart travels along board (m)	Decrease in height of cart (m)	Increase in temperature of cylinder (°C)
1	0.50	0.50	1.8
2	0.58	0.50	1.8
3	0.72	0.50	1.8
4	0.83	0.50	1.8
5	0.97	0.50	1.8

a. In which run was the board vertical?

b. In which run was the board the least steep?

c. Which determines the loss of gravitational potential energy of the cart—the distance traveled along the board or the decrease in height?

21. Ten students whose combined weight is 7,000 N ride down 30 m (about 10 floors) in an elevator. How much thermal energy is produced in the elevator system by the loss in gravitational potential energy of the students?

22. An object with mass m is moving at speed v. How is the kinetic energy of the object affected under each condition?

    a. m is doubled while v remains constant.

    b. v is halved while m remains constant.

    c. Both m and v are doubled.

    d. m is doubled and v is halved.

    e. m is halved and v is doubled.

23. Suppose the friction in the wheels of the cart described in Problem 20 was so great that it could not be neglected. How would this affect each of the following?

    a. the total change in gravitational potential energy of the cart

    b. the total increase in thermal energy of the wheels and the aluminum cylinder as the cart travels down the slanted board

    c. the change in temperature of the aluminum cylinder.

24. To quickly stop a car moving on a horizontal road, you must apply the brakes. To keep a car at a constant speed while it is traveling down a hill, you also must apply the brakes.

    Consider two identical cars of mass 1,000 kg. One, moving horizontally, is brought to a stop from an initial speed of 25 m/s (about 55 mi/h). The other travels down a 300-m-high hill (about 1,000 ft) at constant speed and is not stopped. For which car will the increase in thermal energy of the brakes be greater? (A 1,000-kg car weighs about 10,000 N.)

25. Suppose that a 110-lb student slowly walks down three floors in a building (a total height of about 10 m.)

    a. What is the student's loss of gravitational potential energy?

    b. If this loss could be converted into a gain of thermal energy by the aluminum cylinder used in Experiment 7.1, what would be the rise in temperature of the cylinder?

    c. Why would the student not feel any rise in temperature?

26. Suppose a pendulum is hung from a ceiling. When the pendulum is pulled to one side and released, it swings back and forth.

    a. When is the gravitational potential energy of the pendulum increasing? When is it decreasing?

    b. When is the kinetic energy of the pendulum increasing? When is it decreasing?

THEME FOR A SHORT ESSAY

The word "potential" is used both in everyday language and as an adjective describing a form of energy. Discuss the similarities and the differences between the two uses.

Proportionality

Consider the following ratios:

$$\frac{8}{2}, \ \frac{16}{4}, \ \frac{40}{10}, \ \text{and} \ \frac{48}{12}$$

You can see by inspection or by simplifying that all these ratios equal 4. This fact can also be expressed by the following equations:

$$8 = 4 \times 2, \quad 16 = 4 \times 4, \quad 40 = 4 \times 10, \quad \text{and} \quad 48 = 4 \times 12.$$

Think of the factor 4 in each equation as the number of sides of a square, and the other factor as the length of each side of a square. The product is then equal to the perimeter of a square. This relationship can be modeled with a table (Table A1.1).

The relationship between the length of a side and the perimeter of a square also can be modeled with a formula:

$$\text{Perimeter of square} = 4 \cdot (\text{length of one side}).$$

The data listed in Table A1.1 can also be represented by a graph (Figure A1.1). Notice that the graph is a straight line through the origin.

Figure A1.1

Table A1.1 Perimeter of a Square

Length of one side	Perimeter
2	8
4	16
10	40
12	48
⋮	⋮

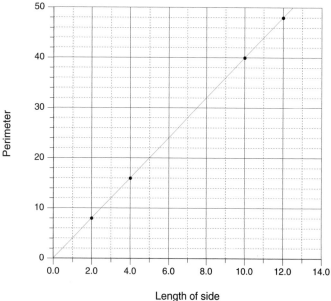

There are many situations in which one quantity equals a constant times another quantity:

$$\text{Quantity 2} = \text{constant} \cdot (\text{quantity 1}) .$$

We call such a relationship a *proportionality*. The constant factor is the *proportionality constant*. Two quantities are proportional to each other when their ratio remains constant. This means that when one quantity is doubled, so is the other. In fact, when one quantity is tripled or multiplied by one-fourth or any other factor, so is the other. Also, the graph of any proportionality is a straight line through the origin.

Usually, the two proportional quantities have units. In our example, both side length and perimeter have units of length, such as meters or centimeters. Because the constant of proportionality is the ratio of the two quantities, its *units* will be the ratio of the *units* of the two quantities. In our example, length of one side and perimeter of a square have the same units. Thus, their ratio is a number without units.

The nature of the proportionality constant is quite different in the case of the stretch of a spring. The data in Table A1.2 were collected by placing different weights on a spring scale. The force (weight) is given in newtons and the stretch in centimeters.

Table A1.2 Stretch of a Spring

Force (N)	Stretch (cm)
0	0
0.45	0.6
1.35	1.8
1.80	2.4
2.25	3.0
3.15	4.2

The graph of stretch as a function of force in Figure A1.2 is indeed a straight line through the origin. Thus, the relationship is a proportionality. You can find the proportionality constant by dividing the vertical coordinate of any point on the graph by its horizontal coordinate. For example, for a force of 2.00 N, the stretch is 2.67 cm. Hence, the proportionality constant is

$$\frac{2.67 \text{ cm}}{2.00 \text{ N}} = 1.34 \text{ cm} / \text{N}$$

In Chapters 6 and 7, you will encounter situations in which a quantity is independently proportional to each of two variables. Here is a familiar example. Suppose you want to cover a rectangular area with some floor covering. For a given width, the cost of the covering will be proportional to the length of the rectangle. By the same reasoning, for a given length, the cost of the covering will be proportional to the width of the rectangle.

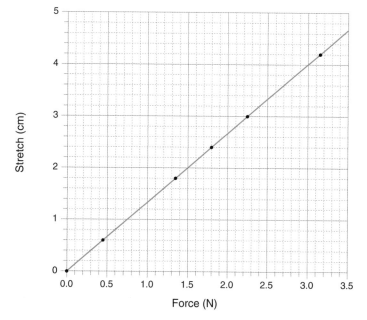

$$\text{Cost} = \text{constant} \cdot \text{length} \quad \text{(for a given width)}$$

$$\text{Cost} = \text{constant} \cdot \text{width} \quad \text{(for a given length)}$$

You already know that the cost of the floor covering is proportional to the area to be covered. But the area of a rectangle equals the product of length and width. Hence, the cost of the floor covering is proportional to the product of length and width:

$$\text{Cost} = \text{constant} \cdot \text{length} \cdot \text{width}.$$

This result can be generalized: when a quantity is independently proportional to each of two variables, it is proportional to their product.

Appendix 2
Graphing

Several considerations enter into constructing a useful graph. Whether you draw the graph yourself or use a computer to do it, you must still make the crucial decisions.

The first step in drawing a graph is to determine which quantity to place on the horizontal axis and which to place on the vertical axis. In Experiment 1.2, Weight and the Spring Scale, you controlled how much weight you added to the container hanging from the spring scale. This makes weight the *independent variable*. The independent variable is on the horizontal axis. The amount of stretch depended on the weight, so it is the *dependent variable*. The dependent variable is on the vertical axis.

The next step is to set suitable *scales* for the axes. Let us begin with the horizontal axis. In Experiment 1.2, you changed the weight by known amounts that ranged from zero to five units. So the scale on the horizontal axis should extend from zero to at least five units.

Creating smaller divisions between units will make it easier to read the graph for weights that are not a whole number of units. However, how the units are divided can be important. For example, for a weight of 2.4 units, reading the amount of stretch is easy when the intervals between consecutive whole numbers are divided into 5 or 10 parts but quite difficult when there are 3 small divisions. To convince yourself of this, try it both ways.

Now consider the vertical axis. Suppose the longest stretch is 37.7 mm. Ending the vertical axis at 37.7 would make it very difficult to produce a useful scale. Extending the scale to 40 mm makes it much simpler.

These guidelines will help you choose suitable scales for your graphs:

1. The scales must cover the entire range of your data points.
2. The scales need not begin at the least value and end at the greatest.
3. Design the scale divisions to make plotting and reading easy.

Once all the points have been plotted, you can draw a smooth curve that passes through, or close to, the points. In Figure 1.3 the smooth curve is a straight line. This line does not pass through all the plotted points. A curve drawn through all the points would contain many little "wiggles" resulting, in all likelihood, from uncertainties in the measurements. Therefore, a smooth curve that is close to most points and shows the general trend is a better picture of the results than a curve that passes through all the points. In some cases, the curve that best fits the data may not pass through any of the data points.

140 APPENDIX 2/GRAPHING

Appendix 3
Conversion of Units

If the world used only one standard set of units of measurement, this Appendix might not be needed. However, for historical reasons as well as for convenience, there are different units in use for almost every imaginable kind of quantity. Therefore, it is often necessary to convert a quantity from one unit to another. To do so, you multiply the quantity by an appropriate conversion factor. A list of often-used conversion factors is given below. Always carry the units through the calculation to be sure that the old units cancel and you are left with the new units.

From	To	Multiply by
mile (mi)	kilometer (km)	$1.61 \; \frac{\text{km}}{\text{mi}}$
inch (in.)	centimeter (cm)	$2.54 \; \frac{\text{cm}}{\text{in.}}$
quart (qt)	liter (L)	$0.94 \; \frac{\text{L}}{\text{qt}}$
pound (lb)	newton (N)	$4.45 \; \frac{\text{N}}{\text{lb}}$
calorie (cal)	joule (J)	$4.18 \; \frac{\text{J}}{\text{cal}}$
inch^2	cm^2	$6.45 \; \frac{\text{cm}^2}{\text{in.}^2}$
lb/in.^{2}	N/cm^2	$0.69 \; \frac{\text{N}/\text{cm}^2}{\text{lb}/\text{in.}^2}$
miles/hour (mi/h)	meters/second (m/s)	$0.45 \; \frac{\text{m}/\text{s}}{\text{mi}/\text{h}}$

Peps   lbs   2.378

To convert in the opposite direction, multiply by the reciprocal of the conversion factor, including the units. For example, to convert from kilometers to miles, multiply by the reciprocal of 1.61 km/mi, or 0.62 mi/km.

Appendix 4
Histograms

In many experiments in this course, you will draw conclusions from data collected by the entire class. However, looking at 15 to 30 different results can be confusing. Graphs often help. We shall illustrate how to construct a suitable graph with an example related to Experiment 1.2, Weight and the Spring Scale. To make the example applicable to other situations as well, we used data measured to the nearest tenth of a unit (Table A4.1).

Table A4.1 Class Data for Experiment 1.2, Weight and the Spring Scale

Team number	Weight of object	Team number	Weight of object
1	1.5	13	1.5
2	2.0	14	2.0
3	1.9	15	1.0
4	2.3	16	2.0
5	1.5	17	1.5
6	2.0	18	1.7
7	1.4	19	1.4
8	2.0	20	2.0
9	1.5	21	1.4
10	1.9	22	2.0
11	1.5	23	1.7
12	1.2	24	1.9

Is there any pattern in the data? Are the results all different or do they cluster around one or two values? An examination of the data reveals that there are a number of values near 1.5 and another set near 2.0. The following graphing procedure will give us a clearer view of this pattern.

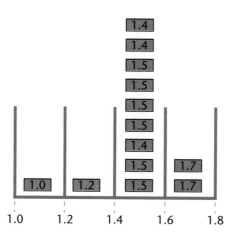

Figure A4.1
Some of the "bins" for depositing data. The data are in the process of being placed in the bins. The numbers under the bins show the bin boundaries.

Imagine that we have some bins, or boxes, marked "1.0 to 1.2," "1.2 to 1.4," and so on. Each team should write its result for the weight on a card and deposit it in the bin that is labeled with the appropriate interval (Figure A4.1). What should we do with the values that are on the boundary between two bins? The simplest solution is always to put them in the bin to the right, as was done in Figure A4.1.

The size (width) of bins (intervals) is important. If the bins are too wide, most of the data will fall into a few bins and the result will be of little value in understanding the data. If the bins are too narrow, the data will be spread out across a large number of bins in an almost flat pattern. Furthermore, it makes no sense to make the bins smaller than the uncertainty of the measurements. For example, if, in Experiment 1.2, you estimated the weight to the nearest 0.5 unit, the bins should be at least 0.5 units wide. In Table A4.1, the weights were estimated to the nearest 0.1 unit. Such an estimate may easily be in error by 0.1 in either direction. Thus, a bin size of 0.2 unit is reasonable.

Once the size of the bins has been determined, it is necessary to set a reference value. The reference value can be either at the center of or on the boundary of a bin. In our example, we set the reference value 1.0 as a boundary of a bin. The other boundaries are, therefore, at 1.2, 1.4, and so on to the right. There could also be bins with boundaries at 0.8, 0.6, and so on the left.

We could have chosen the reference value of 1.0 to be at the center of a bin. Then boundaries for a width of 0.2 would have been at 1.1, 1.3, and so on to the right, and 0.9, 0.7, and so on to the left.

If you are looking for a common value, or have reason to believe the class data will converge on a given value, consider "designing" one bin with that value near the center. Notice that 1.5, which is a frequent value in Table A4.1, is in the middle of a bin.

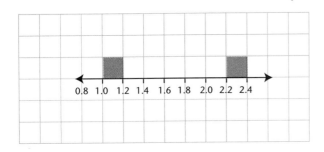

Figure A4.2
The starting point of a histogram: fitting the data on the ends of the horizontal axis.

Once all the cards have been placed in the bins, we count the cards in each bin and plot a bar graph of the results. Of course, we do not need real bins. We can use squares on graph paper instead. As with all graphs, it is a good idea to find the largest and smallest values first in order to decide how to mark the scale on the horizontal axis (Figure A4.2).

Now you can read down Table A4.1 and mark a square for each value in the table. If a value has already appeared, draw its square on top of the one below it. The final result is shown in Figure A4.3. The height of each column tells us the number of measurements that belong in each bin. A plot like Figure A4.3 that presents data by the number of times a value appears in an interval is called a *histogram*.

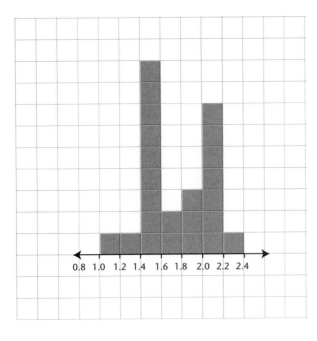

Figure A4.3
A histogram of the data in Table A4.1.

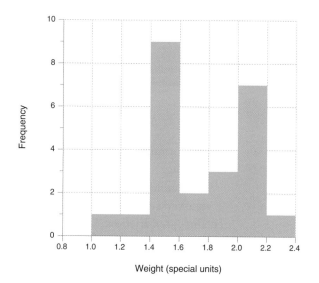

Figure A4.4
A computer-generated histogram of the data in Table A4.1.

A histogram enables us to make some easy observations. Most of the data are bunched around two values—between 1.4 and 1.6, and between 2.0 and 2.2. This is a result of each half of the class using different units of weight.

We can also use computer software to create histograms. Think of drawing a histogram as being divided into two main steps: (1) deciding on the width of the bins and setting the reference value and its location at the center or on a boundary, and (2) placing values in the bins and counting them. The first step requires judgment. The second step is quite mechanical. Making the decisions is best done by you, but the counting and plotting can be left to the computer.

Figure A4.4 shows a histogram of the data in Table A4.1 created by a computer using software written for *FM&E* and *IPS*. Having the computer do the counting and drawing is to your advantage. You can make different histograms quickly and choose the one that best summarizes the data.

Index

H

heat. *See* thermal energy
heat of fusion, 108–109
    definition of, 108
    and melting point, 108
    table, 108
heat of vaporization, 108–109
    and boiling point, 109
    definition of, 108
    table, 109
histograms, constructing, 142–145
Hooke's law, 6–8

I

independent variable, 140
instantaneous speed, 128
internet activity
    Locating an Earthquake, 94

J

joule (J), 101–103
    definition of, 103

K

kg (kilogram), 23
kilogram (kg), 23
kilowatt-hour (kW · h), 111
kinetic energy, 121–126
    definition of, 121
    and speed, 125–126
kW · h (kilowatt-hour), 111

L

law of conservation of energy,
    129–131
lift, 65
lightning, distance from, 88

longitudinal waves, 85–88, 91–94. *See
also* sound
    definition of, 86
    and earthquakes, 91–94
    speed in solids, 91–94
        table, 91

M

magnetic force
    and distance, 8–11
    electromagnet, 19
mass, 22–27
    definition of, 22
    measuring, 22–23
    standard unit for, 23
    and volume, 25–27
    and weight, 22–24
melting point
    definition of, 108
    and heat of vaporization, 108
    table, 108
meniscus, definition of, 26
mercury, in a barometer, 43
motion. *See* force, speed, and kinetic
    energy
motion detector
    for making motion graphs, 72
    for measuring distance, 69–70

N

N (newton), 4–5
net force, 54–56
    determining by
        head-to-tail chain, 55–56
        parallelogram method, 52,55
    as resultant, 55–56
newton (N), 4–5
    definition of, 4
Newton's laws
    first law, 64
    third law, 16–18